T0293070

Elementary Vector Calculus and its Applications with MATLAB Programming

RIVER PUBLISHERS SERIES IN MATHEMATICAL, STATISTICAL AND COMPUTATIONAL MODELLING FOR ENGINEERING

Series Editors:

MANGEY RAM
Graphic Era University, India

TADASHI DOHI
Hiroshima University, Japan

ALIAKBAR MONTAZER HAGHIGHI
Prairie View Texas A&M University, USA

Applied mathematical techniques along with statistical and computational data analysis has become vital skills across the physical sciences. The purpose of this book series is to present novel applications of numerical and computational modelling and data analysis across the applied sciences. We encourage applied mathematicians, statisticians, data scientists and computing engineers working in a comprehensive range of research fields to showcase different techniques and skills, such as differential equations, finite element method, algorithms, discrete mathematics, numerical simulation, machine learning, probability and statistics, fuzzy theory, etc

Books published in the series include professional research monographs, edited volumes, conference proceedings, handbooks and textbooks, which provide new insights for researchers, specialists in industry, and graduate students.

Topics included in this series are as follows:-

- Discrete mathematics and computation
- Fault diagnosis and fault tolerance
- Finite element method (FEM) modeling/simulation
- Fuzzy and possibility theory
- Fuzzy logic and neuro-fuzzy systems for relevant engineering applications
- Game Theory
- Mathematical concepts and applications
- Modelling in engineering applications
- Numerical simulations
- Optimization and algorithms
- Queueing systems
- Resilience
- Stochastic modelling and statistical inference
- Stochastic Processes
- Structural Mechanics
- Theoretical and applied mechanics

For a list of other books in this series, visit www.riverpublishers.com

Elementary Vector Calculus and its Applications with MATLAB Programming

Nita H. Shah
Gujarat University, India

Jitendra Panchal
Parul University, India

NEW YORK AND LONDON

Published 2023 by River Publishers
River Publishers
Alsbjergvej 10, 9260 Gistrup, Denmark
www.riverpublishers.com

Distributed exclusively by Routledge
605 Third Avenue, New York, NY 10017, USA
4 Park Square, Milton Park, Abingdon, Oxon OX14 4RN

Elementary Vector Calculus and its Applications with MATLAB Programming / by Nita H. Shah, Jitendra Panchal.

© 2023 River Publishers. All rights reserved. No part of this publication may be reproduced, stored in a retrieval systems, or transmitted in any form or by any means, mechanical, photocopying, recording or otherwise, without prior written permission of the publishers.

Routledge is an imprint of the Taylor & Francis Group, an informa business

ISBN 978-87-7022-387-4 (print)
ISBN 978-10-0082-423-0 (online)
ISBN 978-10-0336-073-5 (ebook master)

While every effort is made to provide dependable information, the publisher, authors, and editors cannot be held responsible for any errors or omissions.

Contents

Preface ix

List of Figures xi

1 Basic Concept of Vectors and Scalars 1

1.1 Introduction and Importance 1
1.2 Representation of Vectors 1
1.3 Position Vector and Vector Components 2
1.4 Modulus or Absolute Value of a Vector 3
1.5 Zero Vector and Unit Vector 4
1.6 Unit Vectors in the Direction of Axes 4
1.7 Representation of a Vector in terms of Unit Vectors 5
1.8 Addition and Subtraction of Vectors 6
1.9 Product of a Vector with a Scalar 6
1.10 Direction of a Vector 7
1.11 Collinear and Coplanar Vectors 8
1.11.1 Collinear Vectors 8
1.11.2 Coplanar Vectors 8
1.12 Geometric Representation of a Vector Sum 8
1.12.1 Law of Parallelogram of Vectors 8
1.12.2 Law of Triangle of Vectors 9
1.12.3 Properties of Addition of Vectors 9
1.12.4 Properties of Scalar Product 10
1.12.5 Expression of Any Vector in Terms of the Vectors Associated with its Initial Point and Terminal Point . 10
1.12.6 Expression of Any Vector in Terms of Position Vectors 11
1.13 Direction Cosines of a Vector 12
1.14 Exercise 26

2 Scalar and Vector Products 29
2.1 Scalar Product, or Dot Product, or Inner Product 29
2.2 The Measure of Angle Between two Vectors and Projections 30
2.2.1 Properties of a Dot Product 30
2.3 Vector Product or Cross Product or Outer Product of Two Vectors 37
2.4 Geometric Interpretation of a Vector Product 38
2.4.1 Properties of a Vector Product 39
2.5 Application of Scalar and Vector Products 45
2.5.1 Work Done by a Force 46
2.5.2 Moment of a Force About a Point 46
2.6 Exercise 52

3 Vector Differential Calculus 55
3.1 Introduction 55
3.2 Vector and Scalar Functions and Fields 55
3.2.1 Scalar Function and Field 56
3.2.2 Vector Function and Field 56
3.2.3 Level Surfaces 56
3.3 Curve and Arc Length 57
3.3.1 Parametric Representation of Curves 57
3.3.2 Curves with Tangent Vector 58
3.3.2.1 Tangent Vector 59
3.3.2.2 Important Concepts 60
3.3.3 Arc Length 61
3.3.3.1 Unit Tangent Vector 61
3.4 Curvature and Torsion 64
3.4.1 Formulas for Curvature and Torsion 67
3.5 Vector Differentiation 70
3.6 Gradient of a Scalar Field and Directional Derivative 73
3.6.1 Gradient of a Scalar Field 73
3.6.1.1 Properties of Gradient 73
3.6.2 Directional Derivative 74
3.6.2.1 Properties of Gradient 75
3.6.3 Equations of Tangent and Normal to the Level Curves 84
3.6.4 Equation of the Tangent Planes and Normal Lines to the Surfaces 85

3.7 Divergence and Curl of a Vector Field 86
3.7.1 Divergence of a Vector Field 86
3.7.1.1 Physical Interpretation of Divergence . . . 86
3.7.2 Curl of a Vector Field 89
3.7.2.1 Physical Interpretation of Curl 89
3.7.3 Formulae for grad, div, curl Involving Operator ∇ . 96
3.7.3.1 Formulae for grad, div, curl Involving Operator ∇ Once 96
3.7.3.2 Formulae for grad, div, curl Involving Operator ∇ Twice 100
3.8 Exercise 104

4 Vector Integral Calculus **111**
4.1 Introduction 111
4.2 Line Integrals 111
4.2.1 Circulation 112
4.2.2 Work Done by a Force 112
4.3 Path Independence of Line Integrals 113
4.3.1 Theorem: Independent of Path 113
4.4 Surface Integrals 122
4.4.1 Flux 123
4.4.2 Evaluation of Surface Integral 123
4.4.2.1 Component form of Surface Integral . . . 124
4.5 Volume Integrals 129
4.5.1 Component Form of Volume Integral 129
4.6 Exercise 131

5 Green's Theorem, Stokes' Theorem, and Gauss' Theorem **135**
5.1 Green's Theorem (in the Plane) 135
5.1.1 Area of the Plane Region 137
5.2 Stokes' Theorem 146
5.3 Gauss' Divergence Theorem 154
5.4 Exercise 163

6 MATLAB Programming **167**
6.1 Basic of MATLAB Programming 167
6.1.1 Basic of MATLAB Programming 167
6.1.1.1 Introductory MATLAB programmes . . . 168
6.1.1.2 Representation of a Vector in MATLAB . 183

6.1.1.3 Representation of a Matrix in MATLAB . 186
6.2 Some Miscellaneous Examples using MATLAB Programming 188

Index **207**

About the Authors **213**

Preface

Vector calculus is an essential language of mathematical physics. Vector calculus plays a vital role in differential geometry, and the study related to partial differential equations is widely used in physics, engineering, fluid flow, electromagnetic fields, and other disciplines. Vector calculus represents physical quantities in two or three-dimensional space, as well as the variations in these quantities.

The machinery of differential geometry, of which vector calculus is a subset, is used to understand most of the analytic results in a more general form. Many topics in the physical sciences can be mathematically studied using vector calculus techniques.

Description of the book:

This book is meant for readers who have a basic understanding of vector calculus. This book is designed to provide accurate information to readers. The language in the book is kept simple so that all readers can easily understand each concept.

This book begins with the introduction of vectors and scalars in chapter 1. Chapter 1 contains essential basic definitions and concepts, vector in terms of unit vectors, geometric representation of vector sum, and direction cosines. The scalar and vector products, measurement of angle and projections, geometric interpretation of a vector product, and their applications are given in chapter 2. In chapter 3, vector and scalar functions and fields, curves, arc length, formulae for curvature and torsion, and its derivation, curl, divergence, and gradient with important properties and physical interpretation, and important results are given in vector differential calculus. Chapter 4 vector integral calculus includes line integrals, circulation, path independence, surface integrals, volume integrals, and its applications like flux and work done by a force are given. In chapter 5, derivation of Green's theorem, Stokes's theorem, and Gauss' divergence theorem are given with various solve examples. MATLAB programming is given in the last chapter 6 includes basic information about MATLAB. Initially, basic examples are given with proper explanation wherever possible that helps readers to understand basic input and output, arithmetic operations, functions, plotting commands available in MATLAB. Variety of solved programs with MATLAB codes along with compiles and debug outputs. So, the reader can run the program using given codes and observe results.

For MATLAB product information, please contact:
The MathWorks, Inc.
3 Apple Hill Drive
Natick, MA 01760-2098 USA
Tel: 508-647-7000
Fax: 508-647-7001
E-mail: info@mathworks.com
Web: www.mathworks.com

List of Figures

Figure 1.1	Represents the geometrical representation of a vector	1
Figure 1.2	Represents a position vector	2
Figure 1.3	Represents equal and negative vectors	3
Figure 1.4	Represents unit vectors in xy-plane	4
Figure 1.5	Represents unit vectors in the direction of x-axis, y-axis, and z-axis	5
Figure 1.6	Represents a vector $\vec{OP}$ in terms of unit vectors	6
Figure 1.7	Represents the product of a vector with a scalar	7
Figure 1.8	Represents the law of parallelogram of vectors	9
Figure 1.9	Represents the law of the triangle of vectors	9
Figure 1.10	Represents any vector in terms of various vectors associated with its endpoints.	11
Figure 1.11	Represents any vector in terms of the position vector	11
Figure 1.12	Represents direction cosines of a vector	12
Figure 1.13	Represents a parallelogram	14
Figure 1.14	Represents a regular hexagon	15
Figure 1.15	Represents a triangle ΔABC	16
Figure 1.16	A space shuttle of 1000 tons weight hangs from two skyscrapers using steel cables	25
Figure 2.1	Represents a scalar or dot product	29
Figure 2.2	Represents a vector or cross product	38
Figure 2.3	Represents the geometric interpretation of a vector or cross product	39
Figure 2.4	Represents work done by a force $\vec{F}$ on a particle	46
Figure 2.5	Represents the moment of a force about a point	47
Figure 3.1	Represents a curve with a tangent vector	59
Figure 3.2	Represents the plane of curvature of the curve	60
Figure 3.3	Represents the arc rate of rotation of binormal	65
Figure 3.4	Represents $\hat{B}$, $\hat{T}$, and $\hat{N}$ orthogonal unit vectors	66

Figure 3.5 Represents the derivative of the vector $\vec{v}(t)$ 71
Figure 3.6 Represents the directional derivative 74
Figure 3.7 Represents the parallelopiped 87
Figure 4.1 Representation of a vector function defined at every point of a curve C 111
Figure 4.2 Representation of a closed curve C 114
Figure 4.3 Representation of parabola $x = y^2$ 115
Figure 4.4 Representation of the rectangle in xy-plane bounded by lines 117
Figure 4.5 Representation of curved surface S and a plane region R 122
Figure 4.6 Representation of the projection of the plane in the first octant 124
Figure 4.7 Representation of the positive octant of the sphere . 126
Figure 4.8 Representation of the sphere 127
Figure 4.9 Representation of the cylinder in positive octant . . 129
Figure 5.1 Represents the region R bounded by the curve C . 135
Figure 5.2 Represents the region R bounded two parabolas . . 138
Figure 5.3 Represents the plane triangle enclosed by given lines 139
Figure 5.4 Represents the rectangle in the xy-plane bounded by lines 140
Figure 5.5 Represents the triangle in the xy-plane bounded by lines 142
Figure 5.6 Represents the region bounded by the parabola and lines 144
Figure 5.7 Represents an open surface bounded by a closed curve C 146
Figure 5.8 Represents the rectangle bounded by the lines . . . 149
Figure 5.9 Represents the surface of a rectangular lamina bounded by the lines 152
Figure 5.10 Represents the boundary of the triangle 153
Figure 5.11 Represents the region bounded by a closed surface S 155
Figure 5.12 Represents the cube 157
Figure 5.13 Represents the cube 159
Figure 5.14 Represents the circle 161

1

Basic Concept of Vectors and Scalars

1.1 Introduction and Importance

The word "Vector" was first given by W. R. Hamilton. In the nineteenth century, Hamilton and Grassmann have formed vector analysis independently. Today, all physical quantities are classified into two different quantities. The physical quantities can be measured directly or indirectly. Some physical quantities are independent of each other or dependent. A quantity that has its value or a magnitude but no direction then it is called scalar quantity or scalar. For example, time, temperature, density, mass, length, power, distance, area, volume, speed, work, energy, electric charge, frequency, gravitational potential, etc., in this list of examples all quantities are having magnitudes but are independent of the direction. Whereas a quantity that has magnitude, as well as direction, is known as vector quantity or vector. Velocity, acceleration, magnetic field, force, momentum, lift, drag, thrust, displacement, fluid flow, the intensity of an electrical field, centrifugal force, etc., are examples of vector quantities. Vectors are generally denoted by capital bold letters or letters with an arrow-like $\vec{A}$, **A**, or a.

1.2 Representation of Vectors

A geometrical representation of a vector is given in Figure 1.1.

Let O be any arbitrary point in the space and let M be any point in the space. A directed line segment joining both the points is known as the vector $\overrightarrow{OM}$. The length of the vector $\left|\overrightarrow{OM}\right|$ is the magnitude of the vector

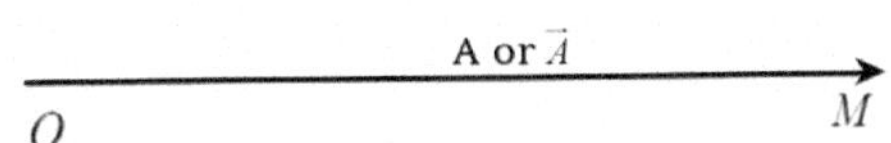

Figure 1.1 Represents the geometrical representation of a vector

$\overrightarrow{OM}$ and is denoted by $|\overrightarrow{OM}|$ or$|\boldsymbol{OM}|$. The point O is called the initial point and the point M is called the terminal point of the vector $|\overrightarrow{OM}|$.

1.3 Position Vector and Vector Components

Consider the cartesian coordinate system shown in Figure 1.2. Let P be any point in the three-dimensional system and let O be the origin then $\overrightarrow{OP}$ is the position vector of the point P. If the vector $\overrightarrow{OP}$ is denoted by $\vec{a}$ then the point is denoted by $P(\vec{a})$. All three axis are perpendicular to each other. The position vector can be obtained by taking perpendiculars on each axis. In the cartesian coordinate system, $AN(= OB)$ represents x-coordinate of P, $BN(= OA)$ represents y-coordinate of P, and PN represents Z-coordinate of P. Thus, the point P is denoted by $P(x, y, z)$. Where x, y, and z are also known as **components** in the direction of the X-axis, Y-axis, and Z-axis respectively.

Equal Vectors: Two vectors with the same direction and magnitude are called equal vectors irrespective of the position of their initial points. In Figure 1.3, $\vec{a}$ and $\vec{b}$ are equal vectors.

Negative Vectors: Two vectors with the same magnitude but opposite in direction are called negative vectors. In Figure 1.3, $\vec{c}$ is a negative vector for both $\vec{a}$ and $\vec{b}$.

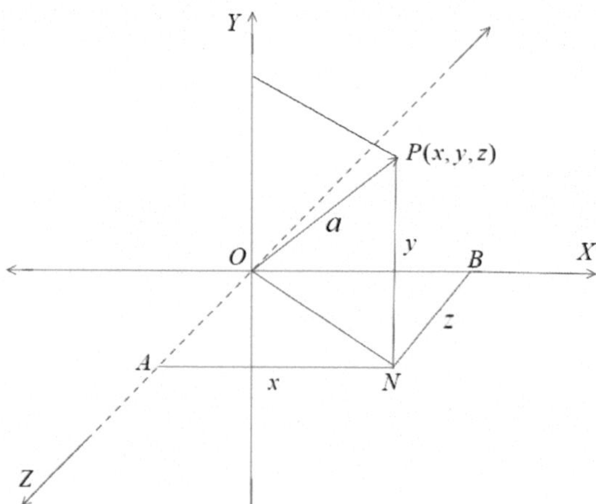

Figure 1.2 Represents a position vector

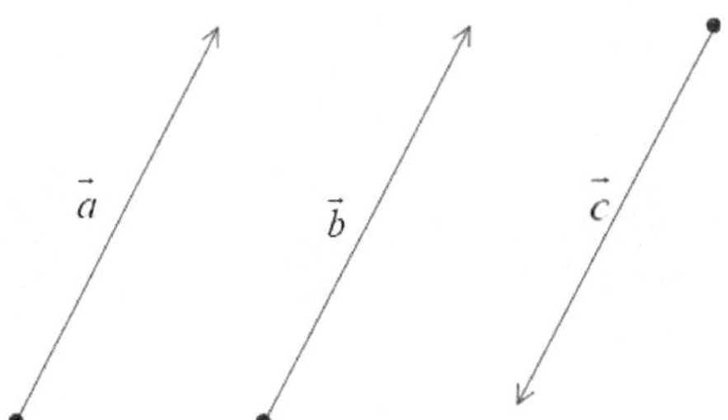

Figure 1.3 Represents equal and negative vectors

1.4 Modulus or Absolute Value of a Vector

In Figure 1.2, $\overrightarrow{OP}$ is the position vector of the point $P(x, y, z)$. The modulus or absolute value of a vector $\overrightarrow{OP}$ is the length of the vector $\overrightarrow{OP}$.

i.e., $OP^2 = BM^2 + MA^2 + PM^2 = x^2 + y^2 + z^2$

$$\therefore \left|\overrightarrow{OP}\right| = OP = \sqrt{x^2 + y^2 + z^2}$$

Illustration 1.1: Find the modulus of the vector $(-3,\ 4, -5)$.

Solution: Let $\overrightarrow{a} = (-3,\ 4, -5)$ be a given vector. It is a three-dimensional vector. Here, $x = -3, y = 4$, and $z = -5$ then the modulus of the vector $\overrightarrow{a}$ is given by

$$|\overrightarrow{a}| = \sqrt{x^2 + y^2 + z^2} = \sqrt{(-3)^2 + (4)^2 + (-5)^2}$$
$$= \sqrt{9 + 16 + 25} = \sqrt{50}$$
$$= 5\sqrt{2}$$

Thus, the modulus of a vector $\overrightarrow{a}$ is $5\sqrt{2}$.

Illustration 1.2: Find the modulus of the vector $(6,\ 8)$.

Solution: Let $\overrightarrow{a} = (6,\ 8)$ be a given vector. It is a two-dimensional vector. Here, $x = 6$ and $y = 8$ then the modulus of the vector $\overrightarrow{a}$ is given by

$$|\overrightarrow{a}| = \sqrt{x^2 + y^2} = \sqrt{(6)^2 + (8)^2}$$
$$= \sqrt{36 + 64} = \sqrt{100}$$
$$= 10$$

Thus, the modulus of a vector $\overrightarrow{a}$ is 10.

1.5 Zero Vector and Unit Vector

Zero Vector: A vector with modulus zero is called a zero vector. It is denoted by 0 or θ. Here, $\theta = 0 = (0, 0, 0)$. So, $|\theta| = |0| = \sqrt{0^2 + 0^2 + 0^2} = 0$. Note that $|\theta| = 0$. $\overrightarrow{AA}$, $\overrightarrow{BB}$ etc. are zero vectors.

Unit Vector: A vector with modulus unity (i.e., 1) is called a unit vector.

Illustration 1.3: $(-1, 0, 0)$ is a unit vector as its modulus is $\sqrt{(-1)^2 + 0^2 + 0^2} = 1$.

Illustration 1.4: $\left(\frac{1}{\sqrt{3}},\ 0,\ -\sqrt{\frac{2}{3}}\right)$ is a unit vector as its modulus is

$$\sqrt{\left(\frac{1}{\sqrt{3}}\right)^2 + 0^2 + \left(-\sqrt{\frac{2}{3}}\right)^2} = \sqrt{\frac{1}{3} + 0 + \frac{2}{3}} = \sqrt{\frac{1+2}{3}} = \sqrt{\frac{3}{3}} = 1.$$

1.6 Unit Vectors in the Direction of Axes

Figure 1.4 represents a two-dimensional cartesian coordinate system (i.e., xy-plane), in which $(1,\ 0)$ and $(0,\ 1)$ are called unit vectors in the direction of x and y axis respectively. The unit vector in the direction of x-coordinate is denoted by $\boldsymbol{i} = (1, 0)$ and in the direction of y-coordinate is denoted by $\boldsymbol{j} = (0, 1)$.

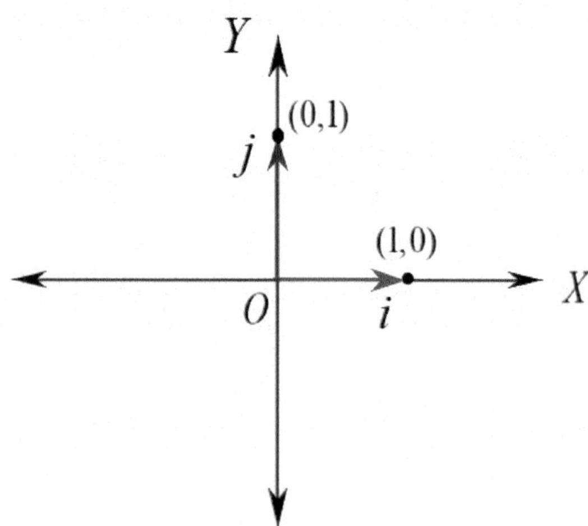

Figure 1.4 Represents unit vectors in xy-plane

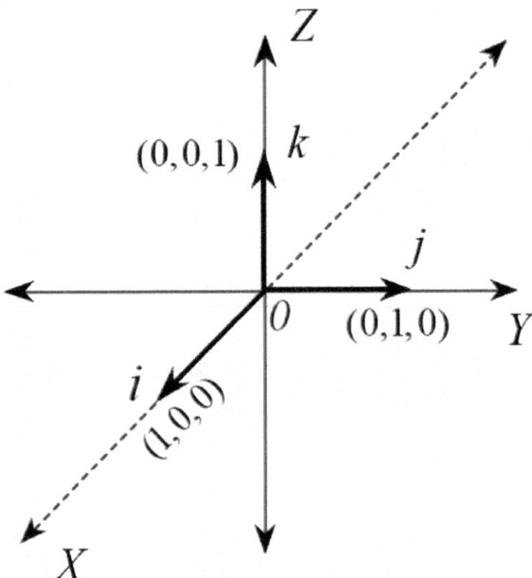

Figure 1.5 Represents unit vectors in the direction of x-axis, y-axis, and z-axis

Figure 1.5 represents a three-dimensional cartesian coordinate system (i.e., xyz-axes), in which $(1,\ 0,\ 0)$, $(0,\ 1,\ 0)$, and $(0,\ 0,\ 1)$ are called unit vectors in the direction of x, y, and z axes respectively. The unit vector in the direction of x-coordinate is denoted by $\boldsymbol{i} = (1, 0, 0)$, in the direction of y-coordinate is denoted by $\boldsymbol{j} = (1, 0, 0)$, and in the direction of z-coordinate is denoted by $\boldsymbol{k} = (1, 0, 0)$. Note that the modulus of each unit vector in the direction of each axis is unity.

1.7 Representation of a Vector in terms of Unit Vectors

Let $P(x, y, z)$ be a vector and $\boldsymbol{i},\ \boldsymbol{j}$, and $\boldsymbol{k}$ be the unit vectors in the direction of $\overrightarrow{OX}$, $\overrightarrow{OY}$ and $\overrightarrow{OZ}$ respectively in the given Figure 1.6.

Consider a perpendicular PM to the plane XOY. From Figure 1.6, we can observe that $ML \perp OX$ and $MN \perp OY$. Then $OL = x$, $ON = y$, and $PM = z$.

$$\therefore \overrightarrow{OL} = x\boldsymbol{i},\ \overrightarrow{ON} = y\boldsymbol{j},\ \text{ and } \overrightarrow{OM} = z\boldsymbol{k}$$

The vectors $x\boldsymbol{i},\ y\boldsymbol{j}$, and $z\boldsymbol{k}$ are called the rectangular components of the vector $\overrightarrow{OP}$ and we can write $\overrightarrow{OP} = x\boldsymbol{i} + y\boldsymbol{j} + z\boldsymbol{k}$. The magnitude of $\overrightarrow{OP} = |\overrightarrow{OP}| = \sqrt{x^2 + y^2 + z^2}$. Thus, $P(x, y, z)$ and $\overrightarrow{OP} = x\boldsymbol{i} + y\boldsymbol{j} + z\boldsymbol{k}$ represent the same vector.

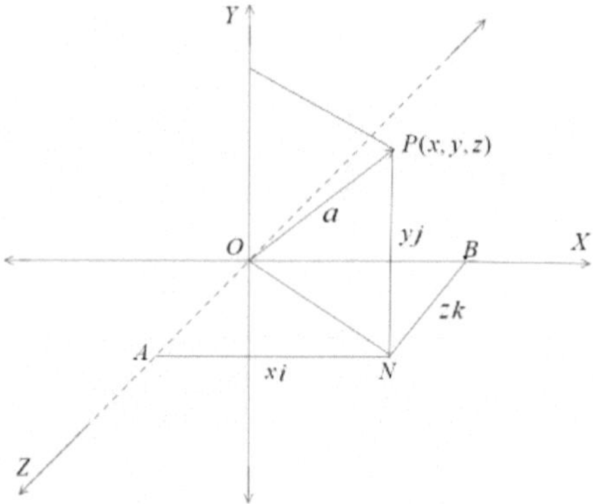

Figure 1.6 Represents a vector $\vec{OP}$ in terms of unit vectors

1.8 Addition and Subtraction of Vectors

If $\vec{x} = (x_1, x_2, x_3)$ and $\vec{y} = (y_1, y_2, y_3)$, then the sum vector $\vec{x} + \vec{y}$ is given by

$$\vec{x} + \vec{y} = (x_1 + y_1, x_2 + y_2, x_3 + y_3)\,.$$

And the subtraction of vector $\vec{x} - \vec{y}$ is given by

$$\vec{x} - \vec{y} = (x_1 - y_1, x_2 - y_2, x_3 - y_3)\,.$$

Illustration 1.5: If $\vec{a} = (4, -3, 2)$ and $\vec{b} = (-2, 5, 3)$.then

$$\vec{a} + \vec{b} = (4 - 2, -3 + 5, 2 + 3) = (2, 2, 5)\,.$$

Illustration 1.6: If $\vec{x} = (4, 10, -2)$ and $\vec{y} = (0, 1, -3)$.then

$$\begin{aligned} \vec{x} - \vec{y} &= (4 - 0, 10 - 1, -2 - 3) \\ &= (4 - 0, 10 - 1, -2 + 3) \\ &= (4, 9, 1)\,. \end{aligned}$$

1.9 Product of a Vector with a Scalar

Definition: Let k be a scalar and $\vec{a}$ be a vector. Then $k\vec{a}$ is defined as a vector whose modulus is k times the modulus of the vector $\vec{a}$ and whose direction is the same as that of $\vec{a}$ or opposite to that of $\vec{a}$ according to k is positive or negative (See Figure 1.7).

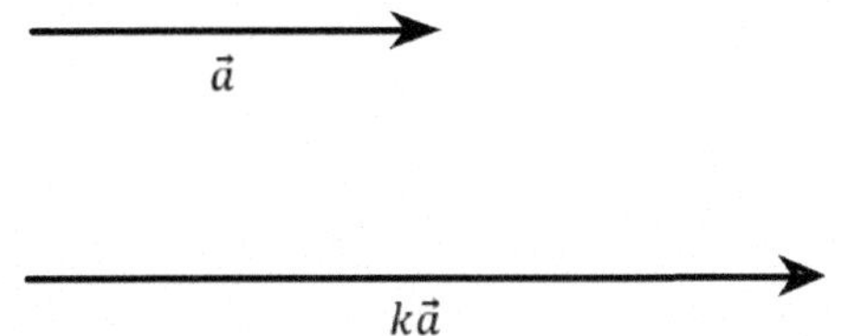

Figure 1.7 Represents the product of a vector with a scalar

Note: $\overrightarrow{a} \parallel \overrightarrow{b} \Leftrightarrow \overrightarrow{a} = k\overrightarrow{b}$ or $\overrightarrow{b} = k\overrightarrow{a}, k \in R$

Illustration 1.7: If $\overrightarrow{a} = (5, -3, 2)$, then

$$3\overrightarrow{a} = 3(5, -3, 2) = (15, -9, 6).$$

Here $3\overrightarrow{a}$ is a vector whose modulus is three times the modulus of $\overrightarrow{a}$ and whose direction is the same as that of $\overrightarrow{a}$.

1.10 Direction of a Vector

In this section, we try to understand the concept of the direction of a vector.

The direction of a vector in the direction from its initial point to its terminal point. Two vectors can be of the same direction, opposite direction, or different directions.

Definition: If $\overrightarrow{a}$ and $\overrightarrow{b}$ are two non-null vectors and

(1) if there exists $\overrightarrow{a}k > 0$ such that $\overrightarrow{a} = k\overrightarrow{b}$. then $\overrightarrow{a}$ and $\overrightarrow{b}$ are of the same direction.

(2) if there exists $\overrightarrow{a}k < 0$ such that $\overrightarrow{a} = k\overrightarrow{b}$. then $\overrightarrow{a}$ and $\overrightarrow{b}$ are of opposite direction,

(3) if there does not exist $\overrightarrow{a}k \in R - \{0\}$ such that $\overrightarrow{a} = k\overrightarrow{b}$, then the directions of $\overrightarrow{a}$ and $\overrightarrow{b}$ are different.

Illustration 1.8: Compare the directions of the vectors $\overrightarrow{a} = (2, -5, 3)$ and $\overrightarrow{b} = (4, -10, 6)$.

Solution: Here $\overrightarrow{a} = (2, -5, 3) = \frac{1}{2}(4, -10, 6)$

$= \frac{1}{2}\overrightarrow{b}$ and $\frac{1}{2} > 0$.

$\therefore \overrightarrow{a}$ and $\overrightarrow{b}$ are of the same direction.

Illustration 1.9: Compare the direction of the vectors $\vec{p} = (3, -2, 1)$ and $\vec{q} = (-9, 6, -3)$

Solution: Here $\vec{q} = (-9, 6, -3) = -3\,(3, -2, 1)$
$= -3\vec{p}$ and $-3 < 0$.
$\therefore$ $\vec{p}$ and $\vec{q}$ are of opposite direction.

Illustration 1.10: Compare the directions of $\vec{g} = (2, 5, 7)$ and $\vec{h} = (3, 1, 6)$.

Solution: Here $\frac{2}{3} \neq \frac{5}{1} \neq \frac{7}{6}$ i.e., there does not exist the same ratio between the elements of $\vec{g}$ and $\vec{h}$.
$\therefore$ We cannot express $\vec{g}$ and $\vec{h}$ in the form $\vec{g} = k\,\vec{h}$. $(k \neq 0)$
$\therefore$The directions of $\vec{g}$ and $\vec{h}$ are different.

1.11 Collinear and Coplanar Vectors

1.11.1 Collinear Vectors

If two vectors $\vec{a}$ and $\vec{b}$ are such that $\vec{a} = k\,\vec{b}$, or $\vec{b} = k\vec{a}$, where $k \in R - \{0\}$, then $\vec{a}$ and $\vec{b}$ are called collinear vectors. Thus $\vec{a}$, $\frac{1}{2}\vec{a}$, $3\vec{a}$, $-\frac{5}{2}\vec{a}$ are collinear vectors.

Collinear vectors can be represented by parallel lines or line segments of the same line.

1.11.2 Coplanar Vectors

Any number of vectors, which are parallel to the same plane, are called coplanar vectors.

If two non-linear vectors $\vec{a}$ and $\vec{b}$ are coplanar, then vector $\vec{R}$ given by $\vec{R} = x\,\vec{a} + y\,\vec{b}$ is coplanar with the vectors $\vec{a}$ and $\vec{b}$ for any $x, y \in R$.

1.12 Geometric Representation of a Vector Sum

1.12.1 Law of Parallelogram of Vectors

If $\vec{a}$ is a vector represented by $\overrightarrow{AB}$ and $\vec{b}$ is a vector represented by $\overrightarrow{AD}$ (the two vectors $\vec{a}$ and $\vec{b}$ have a common initial point), then their sum $\vec{a} + \vec{b}$ is represented in magnitude and direction by $\overrightarrow{AC}$, where $ABCD$ is

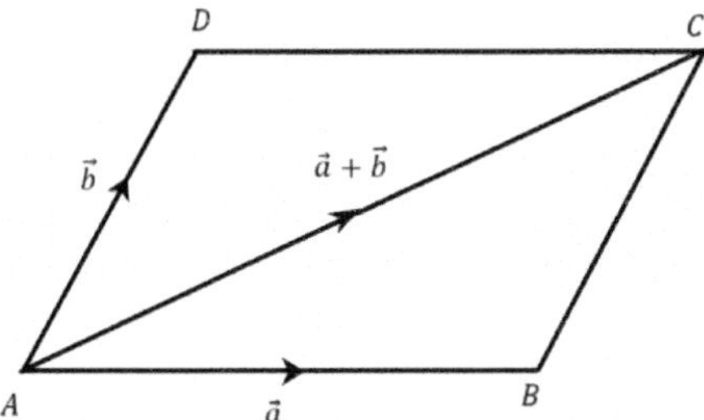

Figure 1.8 Represents the law of parallelogram of vectors

a parallelogram (See Figure 1.8). This method of addition is called the law of the parallelogram of vectors.

1.12.2 Law of Triangle of Vectors

If two vectors $\overrightarrow{u}$ and $\overrightarrow{v}$ are represented by the sides PQ and QR of ΔPQR, then the sum $\overrightarrow{u}+\overrightarrow{v}$ is represented by PR. The direction of $\overrightarrow{u}+\overrightarrow{v}$ is from P to R and its modulus is the length of the side PR of ΔPQR.

This method of addition is called the law of the triangle of vectors.

Here the terminal point of the vector $\overrightarrow{u}$ should be the initial point of the vector $\overrightarrow{v}$. The resultant (sum) vector $\overrightarrow{u}+\overrightarrow{v}$ can be obtained by joining the initial point of the vector $\overrightarrow{u}$ and the terminal point of the vector $\overrightarrow{v}$ as shown in Figure 1.9.

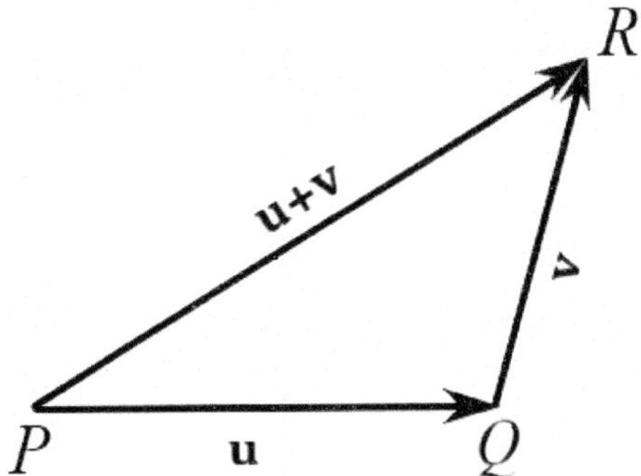

Figure 1.9 Represents the law of the triangle of vectors

1.12.3 Properties of Addition of Vectors

Let $\overrightarrow{a}$, $\overrightarrow{b}$, and $\overrightarrow{c}$ be vectors then

(1) Commutative law: $\vec{a} + \vec{b} = \vec{b} + \vec{a}$

(2) Associative law: $\left(\vec{a} + \vec{b}\right) + \vec{c} = \vec{a} + \left(\vec{b} + \vec{c}\right)$

(3) Identity vector for addition: $\vec{\theta} = (0, 0, 0)$ is identity vector for addition. $\vec{a} + \vec{\theta} = \vec{a} = \vec{\theta} + \vec{a}$

(4) Opposite vector or negative of a vector: For a vector, there exists a vector $(-\vec{a})$ such that $\vec{a} + (-\vec{a}) = \vec{\theta}$. $-\vec{a}$ is called the opposite or negative of $\vec{a}$.

Note: If $\vec{a} = (x, y, z)$, then $-\vec{a} = (-x, -y, -z)$. And $-\vec{a} = -1 \cdot \vec{a}$.

The moduli of $\vec{a}$ and $(-\vec{a})$ are equal but their directions are opposite to each other. We can define the difference of vectors $\vec{a}$ and $\vec{b}$ as the sum of $\vec{a}$ and $\left(-\vec{b}\right)$ i.e. $\vec{a} - \vec{b} = \vec{a} + \left(-\vec{b}\right)$.

1.12.4 Properties of Scalar Product

Let $\vec{a}, \vec{b}, \vec{c}$ be vectors and $m, n \in R$ be scalars.

(1) $m\vec{a} = \vec{a}m$

(2) $m\left(n\vec{a}\right) = n\left(m\vec{a}\right) = \left(mn\vec{a}\right)$

(3) Distributive law: $(m + n)\vec{a} = m\vec{a} + n\vec{a}$

(4) $m\left(\vec{a} + \vec{b}\right) = m\vec{a} + m\vec{b}$

(5) $\theta\vec{a} = \vec{a}\theta = \vec{\theta}$

1.12.5 Expression of Any Vector in Terms of the Vectors Associated with its Initial Point and Terminal Point

Let $\overrightarrow{BC}$ be a vector. Take a point A which is not on $\overleftrightarrow{BC}$ join AB and AC. We now have three different vectors $\overrightarrow{AB}$, $\overrightarrow{AC}$, and $\overrightarrow{BC}$ (See Figure 1.10). By the law of the triangle of vectors, we have

$$\overrightarrow{AB} + \overrightarrow{BC} = \overrightarrow{AC}$$

$$\therefore \overrightarrow{BC} = \overrightarrow{AC} - \overrightarrow{AB}$$

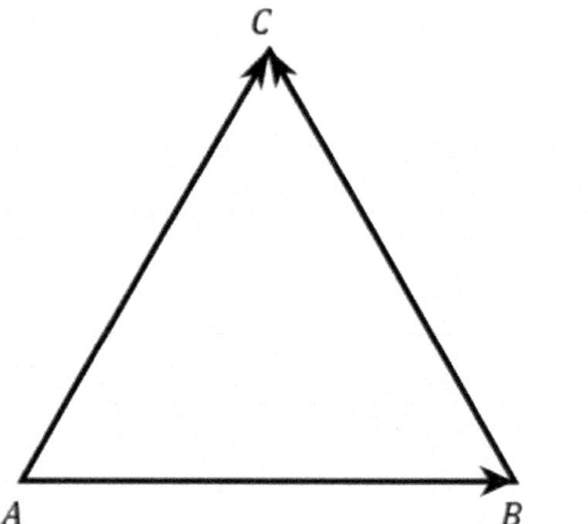

Figure 1.10 Represents any vector in terms of various vectors associated with its endpoints.

In other words, $\overrightarrow{BC}$= vector of point $C-$ vector of point B.

In general, any vector=vector of its terminal point-vector of its initial point.

1.12.6 Expression of Any Vector in Terms of Position Vectors

The method of expression of a vector discussed above is true for any point A. We can take the origin O in place of the point A.

As shown in Figure 1.11,

$$\overrightarrow{OA}+\overrightarrow{AB}=\overrightarrow{OB}$$

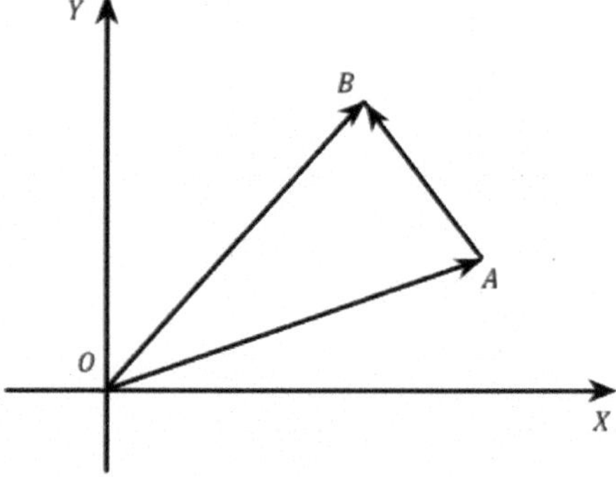

Figure 1.11 Represents any vector in terms of the position vector

$$\therefore \overrightarrow{AB} = \overrightarrow{OB} - \overrightarrow{OA}$$

In other words, any vector=position vector of its terminal point –position vector of its terminal point –position vector of its initial point.

Illustration 1.11: Let the position vectors of the points A and B are (2, 5, -3) and (3, -2, 5) respectively. Then

$AB =$ position vector of $B -$ position vector of A

$$= (3, -2, 5) - (2, 5, -3) = (1, -7, 8)$$

1.13 Direction Cosines of a Vector

In two dimensions, a vector makes angles with only two axes, namely X-axis and Y-axis. Hence it is easier to understand. But in three-dimensional space, a vector makes angles with three axes, and to understand the position of the vector, direction cosines of the vector are useful. In Figure 1.12, three important angles are shown considering OP.

(i) Angle formed by OP with the X-axis is $\angle POX$. It is denoted by α.

(ii) Angle formed by OP with the Y-axis is $\angle POY$. It is denoted by β.

(iii) Angle formed by OP with the Z-axis is $\angle POZ$. It is denoted by γ.

The cosines of these angles are called direction cosines. Thus, we get three direction cosines.

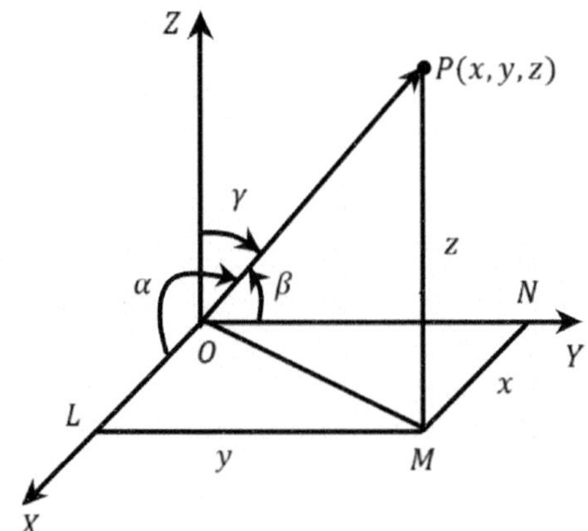

Figure 1.12 Represents direction cosines of a vector

(1) $l = \cos\alpha \; = \frac{x}{OP}$. Taking $OP = r$, we have $\cos a \; = \frac{x}{r}$.
Now, $r^2 = x^2 + y^2 + z^2$.

$$\therefore OP = r = \sqrt{x^2 + y^2 + z^2}$$

$$\therefore l = \cos\alpha \; = \frac{x}{\sqrt{x^2 + y^2 + z^2}} \tag{1.1}$$

(2) $\cos\beta$ is denoted by m. It is connected with the Y-axis. As explained above.

$$m = \cos\beta \; = \frac{y}{\sqrt{x^2 + y^2 + z^2}} \tag{1.2}$$

(3) $\cos\gamma$ is denoted by m. It is connected with the Z-axis.

$$n = \cos\gamma \; = \frac{z}{\sqrt{x^2 + y^2 + z^2}} \tag{1.3}$$

Relation among l, m and n,

Squaring and adding (1.1), (1.2), and (1.3), we get

$$l^2 + m^2 + n^2 = \frac{x^2 + y^2 + z^2}{x^2 + y^2 + z^2} = 1.$$

Thus $l^2 + m^2 + n^2 = 1$

i.e., $\cos^2\alpha \; + \cos^2\beta \; + \cos^2\gamma \; = 1$.

Note that the direction cosines of a vector are the components of its unit vector.

Illustration 1.12: If the position vectors of the vertices A, B, and $C$of the parallelogram $ABCD$ are $\vec{a}$, $\vec{b}$ and $\vec{c}$ respectively, find the position vector of the vertex D.

Solution: Here $OA \; = \; \vec{a}, OB \; = \; \vec{b}$, and $OC \; = \; \vec{c}$. $ABCD$ is a parallelogram (See Figure 1.13).

$$\therefore AD = BC = OC - OB = \vec{c} - \vec{b}$$

Now,

$$\therefore AD = OD - OA$$

$$\therefore OD = AD + OA = \left(\vec{c} - \vec{b}\right) + \vec{a}$$

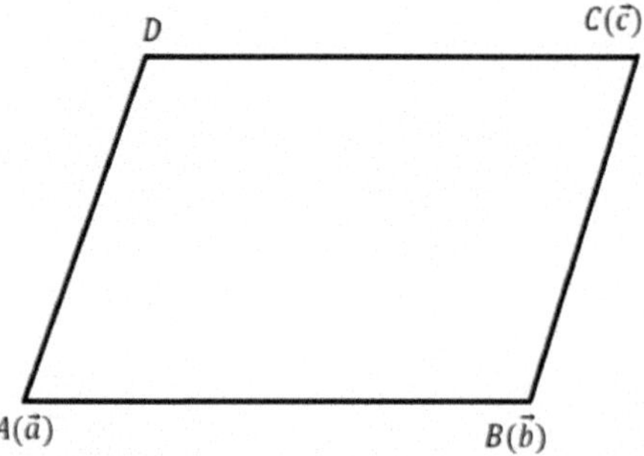

Figure 1.13 Represents a parallelogram

$$= \vec{a} - \vec{b} + \vec{c}$$

Illustration 1.13: If the position vectors of the points A, B, C, D are respectively $\vec{a}, \vec{b}, 3\vec{a} - \vec{b}, -\vec{a} + 3\vec{b}$, then express the vectors AB, AC, BC, BD, and CD in terms of $\vec{a}$ and $\vec{b}$.

Solution: Here $OA = \vec{a}, OB = \vec{b}, OC = 3\vec{a} - \vec{b}$, and $OD = -\vec{a} + 3\vec{b}$.
Now, $AB = OB - OA = \vec{b} - \vec{a}$

$$AC = OC - OA = 3\vec{a} - \vec{b} - \vec{a} = 2\vec{a} - \vec{b}$$

$$AD = OD - OA = -\vec{a} + 3\vec{b} - \vec{a} = 3\vec{b} - 2\vec{a}$$

$$BC = OC - OB = 3\vec{a} - \vec{b} - \vec{b} = 3\vec{a} - 2\vec{b}$$

$$BD = OD - OB = -\vec{a} + 3\vec{b} - \vec{b} = 2\vec{b} - \vec{a}$$

$$CD = OD - OC = -\vec{a} + 3\vec{b} - \left(3\vec{a} - \vec{b}\right)$$

$$= -\vec{a} + 3\vec{b} - 3\vec{a} + \vec{b} = 4\left(\vec{b} - \vec{a}\right)$$

Illustration 1.14: If $ABCDEF$ is a regular hexagon, prove that

$$AB + AC + AD + AE + AF = 3AD.$$

Solution: Suppose $AB = \vec{a}, BC = \vec{b}$, and $CD = \vec{c}$.
As $ABCDEF$ is a regular hexagon,

$$CD = AF = \vec{c},$$

$AB = ED = \vec{a}$ and $BC = FE = \vec{b}$.

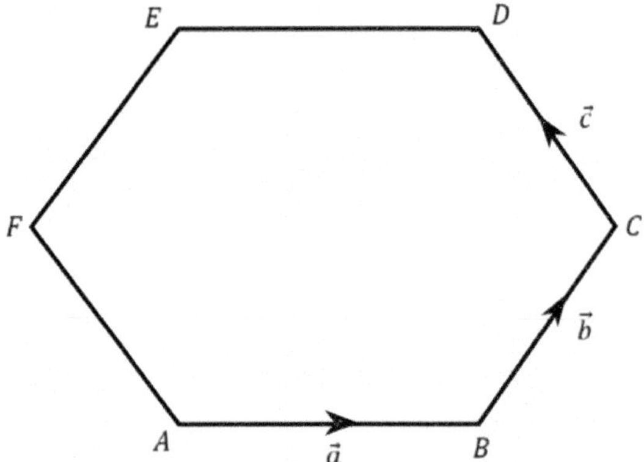

Figure 1.14 Represents a regular hexagon

Now $AB = \overrightarrow{a}$, $AC = AB + BC = \overrightarrow{a} + \overrightarrow{b}$, $AD = AC + CD = \overrightarrow{a} + \overrightarrow{b} + \overrightarrow{c}$

$$\begin{aligned}
&AE = AD + DC = AD - ED\\
&= \overrightarrow{a} + \overrightarrow{b} + \overrightarrow{c} - \overrightarrow{a} = \overrightarrow{b} + \overrightarrow{c}\\
&AF = CD = \overrightarrow{c}\\
&\therefore LHS = AB + AC + AD + AE + AF\\
&= \overrightarrow{a} + \overrightarrow{a} + \overrightarrow{b} + \overrightarrow{a} + \overrightarrow{b} + \overrightarrow{c} + \overrightarrow{b} + \overrightarrow{c} + \overrightarrow{c}\\
&= 3\left(\overrightarrow{a} + \overrightarrow{b} + \overrightarrow{c}\right) = 3AD = RHS
\end{aligned}$$

Another method: From Figure 1.14, we have

$$\begin{aligned}
&LHS = AB + AC + AD + AE + AF\\
&= ED + AC + AD + AE + CD\\
&(\because AB = ED \text{ and } CD = AF)\\
&= (AC + CD) + (AE + ED) + AD\\
&= AD + AD + AD = 3AD = RHS
\end{aligned}$$

Illustration 1.15: If the sides AB and AC of ΔABC (See Figure 1.15) represent two vectors and M is the mid-point of the side BC, then prove that

$$AB + AC = 2AM$$

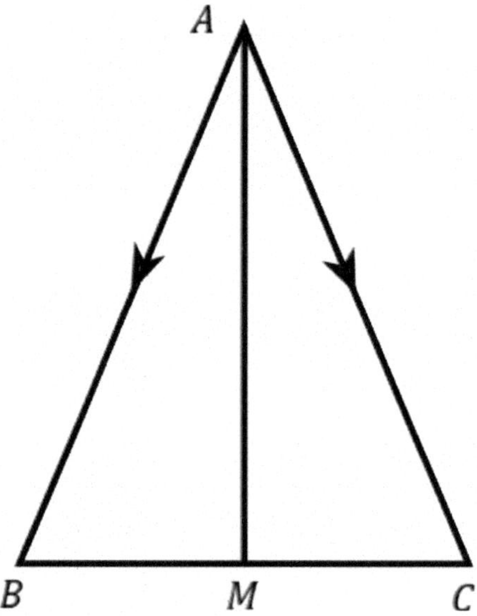

Figure 1.15 Represents a triangle ΔABC

Solution:

Here

$$AB + BM = AM \tag{1.4}$$

and

$$AC + CM = AM \tag{1.5}$$

Adding (1.4) and (1.5), we have

$$AB + BM + AC + CM = 2AM \tag{1.6}$$

But BM and CM are opposite vectors.

$$\therefore BM + CM = 0 \tag{1.7}$$

Using (1.7) in (1.6), we get

$$AB + AC = 2AM$$

Illustration 1.16: Find position vectors, moduli, unit vectors, and direction cosines for vectors represented by the following points:

(i) $P\,(3, -4)$ (ii) $Q\,(6, 2)$ (iii) $R\,(-4, -6)$

Solution:

(i) $OP = \overrightarrow{r} = (3, -4) = 3\hat{\imath} - 4\hat{\jmath}$

Modulus $= |OP| = |\overrightarrow{r}| = \sqrt{(3)^2 + (-4)^2}$

$$= \sqrt{9 + 16} = \sqrt{25} = 5$$

$$\text{Unit vector} \quad \hat{r} = \frac{\overrightarrow{r}}{|\overrightarrow{r}|} = \frac{3\hat{\imath} - 4\hat{\jmath}}{5} = \frac{3}{5}\hat{\imath} - \frac{4}{5}\hat{\jmath}$$

Direction cosines: $l = \frac{3}{5}, m = -\frac{4}{5}$

(ii) $OQ = \overrightarrow{r} = (6, 2) = 6\hat{\imath} + 2\hat{\jmath}$

Modulus $= |\overrightarrow{r}| = \sqrt{6^2 + 2^2} = \sqrt{36 + 4} = \sqrt{40} = 2\sqrt{10}$

$$\text{Unit vector } \hat{r} = \frac{\overrightarrow{r}}{|\overrightarrow{r}|} = \frac{6\hat{\imath} - 2\hat{\jmath}}{2\sqrt{10}} = \frac{3}{\sqrt{10}}\hat{\imath} + \frac{1}{\sqrt{10}}\hat{\jmath}$$

Direction cosines: $l = \frac{3}{\sqrt{10}}, m = \frac{1}{\sqrt{10}}$.

(iii) $OR = \overrightarrow{r} = (-4, -6) = 4\hat{\imath} - 6\hat{\jmath}$

Modulus $= |\overrightarrow{r}| = \sqrt{(-4)^2 + (6)^2} = \sqrt{16 + 36} = \sqrt{52} = 2\sqrt{13}$

$$\text{Unit vector } \hat{r} = \frac{\overrightarrow{r}}{|\overrightarrow{r}|} = \frac{-4\hat{\imath} - 6\hat{\jmath}}{2\sqrt{13}} = \frac{2}{\sqrt{13}}\hat{\imath} - \frac{3}{\sqrt{13}}\hat{\jmath}$$

Direction cosines: $l = -\frac{2}{\sqrt{13}}, m = -\frac{3}{\sqrt{13}}$

Illustration 1.17: If $\overrightarrow{x} = \left(1, \frac{1}{2}\right), \overrightarrow{y} = \left(\frac{1}{\sqrt{2}}, \frac{1}{\sqrt{2}}\right)$ and $\overrightarrow{z} = \left(-2, -\frac{3}{2}\right)$ then

(i) Find a unit vector in the direction of $\overrightarrow{x} + \overrightarrow{z}$,

(ii) Find a unit vector in the direction of $2\overrightarrow{x} - \sqrt{2}\overrightarrow{y} + 2\overrightarrow{z}$.

Solution:

(i) Here $\overrightarrow{x} + \overrightarrow{z} = \left(1, \frac{1}{2}\right) + \left(-2, -\frac{3}{2}\right)$

$$= \left(1 - 2, \frac{1}{2} - \frac{3}{2}\right)$$

$$= (-1, -1) = -\hat{i} - \hat{j}$$

$$\therefore |\vec{x} + \vec{z}| = \sqrt{(-1)^2 + (-1)^2} = \sqrt{1+1} = \sqrt{2}$$

$\therefore$ The unit vector in the direction of $\vec{x} + \vec{z}$

$$= \frac{\vec{x} + \vec{z}}{|\vec{x} + \vec{z}|} = \frac{1}{\sqrt{2}}\left(-\hat{i} - \hat{j}\right) = -\frac{1}{\sqrt{2}}\hat{i} - \frac{1}{\sqrt{2}}\hat{j}$$

$$= \left(-\frac{1}{\sqrt{2}}, -\frac{1}{\sqrt{2}}\right)$$

(ii) $2\vec{x} - \sqrt{2}\vec{y} + 2\vec{z}$

$$= 2\left(1, \frac{1}{2}\right) - \sqrt{2}\left(\frac{1}{\sqrt{2}}, \frac{1}{\sqrt{2}}\right) + 2\left(-2, -\frac{3}{2}\right)$$

$$= (2, 1) - (1, 1) + (-4, -3)$$

$$= (2 - 1 - 4, 1 - 1 - 3) = (-3, -3)$$

$$\left|2\vec{x} - \sqrt{2}\vec{y} + 2\vec{z}\right| = \sqrt{(-3)^2 + (-3)^2}$$

$$= \sqrt{9+9} = \sqrt{18} = 3\sqrt{2}$$

$$\therefore \text{Required unit vector} = \frac{2\vec{x} - \sqrt{2}\vec{y} + 2\vec{z}}{|2\vec{x} - \sqrt{2}\vec{y} + 2\vec{z}|}$$

$$= \frac{1}{3\sqrt{2}}(-3, -3) = \left(-\frac{1}{\sqrt{2}}, -\frac{1}{\sqrt{2}}\right)$$

Illustration 1.18: Let $\vec{a} = (2, -1, 2)$ be a given vector.

(i) Find the unit vector in the direction of $\vec{a}$.

(ii) Find a direction cosine of $\vec{a}$.

(iii) Find a vector of magnitude 6 in the direction of $\vec{a}$.

(iv) Find a vector of magnitude 4 in the opposite direction of $\vec{a}$.

Solution: Here $\vec{a} = (2, -1, 2) = 2\hat{i} - \hat{j} + 2\hat{k}$.

(i) Unit vector in the direction of a

$$\hat{a} = \frac{\vec{a}}{|\vec{a}|} = \frac{2\hat{i} - \hat{j} + 2\hat{k}}{\sqrt{4+1+4}} = \frac{1}{3}\left(2\hat{i} - \hat{j} + 2\hat{k}\right)$$
$$= \frac{2}{3}\hat{i} - \frac{1}{3}\hat{j} + \frac{2}{3}\hat{k}$$

(ii) Direction cosines: $l = \frac{2}{3}, m = -\frac{1}{3}, n = \frac{2}{3}$

(iii) Vector of magnitude 6 in the direction of $\vec{a}$

$$= 6\hat{a} = \frac{6}{3}\left(2\hat{i} - \hat{j} + 2\hat{k}\right) = 4\hat{i} - 2\hat{j} + 4\hat{k}$$

(iv) Vector of magnitude 4 in the opposite direction of $\vec{a}$

$$= -4\hat{a} = -\frac{4}{3}\left(2\hat{i} - \hat{j} + 2\hat{k}\right) = -\frac{8}{3}\hat{i} + \frac{4}{3}\hat{j} - \frac{8}{3}\hat{k}$$

Illustration 1.19: Answer the followings:

(i) If $\vec{a} = 3\hat{i} - \hat{j} - 4\hat{k}$, $\vec{b} = -2\hat{i} + 4\hat{j} - 3\hat{k}$ and $\vec{c} = \hat{i} + 2\hat{j} - \hat{k}$, then find the direction cosines of the vector $3\vec{a} - 2\vec{b} + 4\vec{c}$.

(ii) If $\vec{a} = 2\hat{i} + \hat{j} - \hat{k}$, $\vec{b} = \hat{i} - \hat{j} + 2\hat{k}$ and $\vec{c} = \hat{i} - 2\hat{j} + \hat{k}$, then find the direction cosines of $\vec{a} + \vec{b} - 2\vec{c}$.

(iii) If $\vec{a} = (3, -1, -4)$, $\vec{b} = (-2, 4, -3)$ and $\vec{c} = (-1, 2, -1)$ then find $\left|3\vec{a} - 2\vec{b} + 4\vec{c}\right|$.

(iv) If $\vec{a} = \hat{i} + \hat{j}$, $\vec{b} = \hat{j} + \hat{k}$, and $\vec{c} = \hat{k} + \hat{i}$, then find $\left|2\vec{a} - 3\vec{b} - 5\vec{c}\right|$.

(v) If $\vec{a} = 5\hat{i} - 3\hat{j} + 2\hat{k}$, $\vec{b} = 2\hat{i} + 3\hat{j} - \hat{k}$ and $\vec{c} = \hat{i} + 2\hat{j} + 3\hat{k}$, then find the value of $\left|2\vec{a} - 3\vec{b} + 4\vec{c}\right|$.

(vi) If $\vec{a} = 3\hat{i} - 2\hat{j} + \hat{k}$, $\vec{b} = 2\hat{i} - 4\hat{j} - 3\hat{k}$, and $\vec{c} = -\hat{i} + 2\hat{j} + 2\hat{k}$, find $\left|2\vec{a} - 3\vec{b} - 5\vec{c}\right|$.

(vii) If $\vec{a} = 2\hat{i} + \hat{j} - \hat{k}, \vec{b} = \hat{i} - \hat{j} + 2\hat{k}$ and $\vec{c} = \hat{i} - 2\hat{j} + \hat{k}$ then find $\left|\vec{a} + \vec{b} - 2\vec{c}\right|$.

(viii) If $\vec{a} = (1, 2, 1), \vec{b} = (2, 1, 1)$ and $\vec{c} = (3, 4, 1)$ then find $\left|\vec{a} + 2\vec{b} + \vec{c}\right|$.

(ix) If $\vec{a} = \hat{j} + \hat{k} - \hat{i}$ and $\vec{b} = 2\hat{i} + \hat{j} - 3\hat{k}$ then find $\left|2\vec{a} + 3\vec{b}\right|$.

(x) If $\vec{a} = (1, 2, 1), \vec{b} = (1, -1, 2)$ and $\vec{c} = (3, 2, -1)$ then find $\left|3\vec{a} + \vec{b} - 2\vec{c}\right|$.

Solution:

(i) Here $\vec{a} = (3, -1, -4), \vec{b} = (-2, 4, -3)$ and $\vec{c} = (1, 2, -1)$.

Let $\vec{x} = 3\vec{a} - 2\vec{b} + 4\vec{c}$.

$$\therefore \vec{x} = 3(3, -1, -4) - 2(-2, 4, -3) + (1, 2, -1)$$
$$= (9, -3, -12) - (-2, 4, -3) + (4, 8, -4)$$
$$= (9 + 4 + 4, -3 - 8 + 8, -12 + 6 - 4)$$
$$= (17, -3, -10)$$
$$\therefore |\vec{x}| = \sqrt{17^2 + (-3)^2 + (-10)^2}$$
$$= \sqrt{289 + 9 + 100} = \sqrt{398}$$

If l, m, n are the direction cosines of $\vec{x}$, then

$$l = \frac{x_1}{|x|} = \frac{17}{\sqrt{398}}, m = \frac{x_2}{|x|} = \frac{-3}{\sqrt{398}}, n = \frac{x_3}{|x|} = \frac{-10}{\sqrt{398}}$$

$$\therefore cos\alpha = \frac{17}{\sqrt{398}}, cos\beta = \frac{-8}{\sqrt{398}}, cos\gamma = \frac{-10}{\sqrt{398}}$$

(ii) $$\vec{a} + \vec{b} - 2\vec{c} = 2\hat{i} + \hat{j} - \hat{k} + \hat{i} - \hat{j} + 2\hat{k} - 2\left(\hat{i} - 2\hat{j} + \hat{k}\right)$$
$$= 3\hat{i} + \hat{k} - 2\hat{i} + 4\hat{j} - 2\hat{k}$$

$$= \hat{i} + 4\hat{j} - \hat{k} = (1, 4, -1)$$

$$\therefore |\vec{r}| = \left|\vec{a} + \vec{b} - 2\vec{c}\right| = \sqrt{1^2 + 4^2 + (-1)^2}$$

$$= \sqrt{1 + 16 + 1} = \sqrt{18} = 3\sqrt{2}$$

If l, m, n are the direction cosines of $\vec{a} + \vec{b} - 2\vec{c}$, then

$$l = cos\alpha = \frac{x_1}{|\vec{r}|} = \frac{1}{3\sqrt{2}},\ m = cos\beta = \frac{x_2}{|\vec{r}|} = \frac{4}{3\sqrt{2}},$$

$$n = cos\gamma = \frac{x_3}{|\vec{r}|} = \frac{-1}{3\sqrt{2}}$$

(iii) $3\vec{a} - 2\vec{b} + 4\vec{c}$

$$= 3(3, -1, -4) - 2(-2, 4, -3) + (-1, 2, -1)$$
$$= (9, -3, -12) + (4, -8, 6) + (-4, 8, -4)$$
$$= (9 + 4 - 4, -3 - 8 + 8, -12 + 6 - 4)$$
$$= (9, -3, -10)$$
$$\therefore \left|3\vec{a} - 2\vec{b} + 4\vec{c}\right| = \sqrt{(9)^2 + (-3)^2 + (-10)^2}$$
$$= \sqrt{81 + 9 + 100} = \sqrt{190}$$

(iv) Here $\vec{a} = \hat{i} + \hat{j}, \vec{b} = \hat{j} + \hat{k}, \vec{c} = \hat{k} + \hat{i} = \hat{i} + \hat{k}$

$$\therefore \vec{a} = (1, 1, 0), \vec{b} = (0, 1, 1), \vec{c} = (1, 0, 1)$$
$$\therefore 2\vec{a} = (2, 2, 0), 3\vec{b} = (0, 3, 3), 5\vec{c} = (5, 0, 5)$$
$$\therefore 2\vec{a} - 3\vec{b} - 5\vec{c} = (2, 2, 0) - (0, 3, 3) - (5, 0, 5)$$
$$= (2 - 0 - 5, 2 - 3 - 0, 0 - 3 - 5)$$
$$\therefore 2\vec{a} - 3\vec{b} - 5\vec{c} = (-3, -1, -8)$$
$$\therefore \left|2\vec{a} - 3\vec{b} - 5\vec{c}\right| = \sqrt{(-3)^2 + (-1)^2 + (-8)^2}$$

($\because$ Definition of magnitude)

$$= \sqrt{9 + 1 + 64} = \sqrt{74}$$

(v) Here $\vec{a} = (5, -3, 2)$, $\vec{b} = (2, 3, -1)$, $\vec{c} = (1, 2, 3)$.

$$\text{Let } \vec{x} = 2\vec{a} - 3\vec{b} + 4\vec{c}$$
$$\therefore \vec{x} = 2(5, -3, 2) - 3(2, 3, -1) + 4(1, 2, 3)$$
$$= (10, -6, 4) - (6, 9, -3) + (4, 8, 12)$$
$$= (10 - 6 + 4, -6 - 9 + 8, 4 + 3 + 12)$$
$$\therefore \vec{x} = (8, -7, 19)$$
$$\therefore |\vec{x}| = \sqrt{8^2 + (-7^2) + 19^2}$$
($\because$ Definition of magnitude)
$$= \sqrt{64 + 49 + 361} = \sqrt{474}$$

(vi) Here $\vec{a} = (3, -2, 1)$, $\vec{b} = (2, -4, -3)$, $\vec{c} = (-1, 2, 2)$

$$\therefore 2\vec{a} - 3\vec{b} - 5\vec{c} = 2(3, -2, 1) - 3(2, -4, -3) - 5(-1, 2, 2)$$
$$= (6, -4, 2) + (-6, 12, 9) + (5, -10, -10)$$
$$= (6 - 6 + 5, -4, 12, 10, 2 + 9 - 10)$$
$$\therefore 2\vec{a} - 3\vec{b} - 5\vec{c} = (5, -2, 1)$$
$$\therefore \left|2\vec{a} - 3\vec{b} - 5\vec{c}\right| = \sqrt{(5)^2 + (-2)^2 + (1)^2}$$
($\because$ Definition of magnitude)
$$= \sqrt{25 + 4 + 1} = \sqrt{30}$$

(vii) Given that $\vec{a} = (3, -1, -4)$, $\vec{b} = (-2, 4, -3)$, $\vec{c} = (-1, 2, -5)$

$$\therefore \vec{a} + 2\vec{b} - \vec{c} = 2(3, -1, 4) + 2(-2, 4, -3) - (-1, 2, -5)$$
$$= (3, -1, -4) + (-4, 8, -6) + (1, -2, 5)$$
$$= (3 - 4 + 1, -1 + 8 - 2, -4 - 6 + 5)$$
$$\therefore \vec{a} + 2\vec{b} - \vec{c} = (0, 5, -5)$$
$$\therefore \left|\vec{a} + 2\vec{b} - \vec{c}\right| = \sqrt{0 + 5^2 + (-5)^2}$$
($\because$ Definition of magnitude)
$$= \sqrt{25 + 25} = \sqrt{50}$$
$$\therefore \left|\vec{a} + 2\vec{b} - \vec{c}\right| = 5\sqrt{2}$$

(viii) Here $\overrightarrow{a} = 2\hat{i} + \hat{j} - \hat{k}$,

$$\therefore \overrightarrow{a} = (2, 1, -1),$$
$$\overrightarrow{b} = \hat{i} - \hat{j} + 2\hat{k}$$
$$\therefore \overrightarrow{b} = (1, -1, 2) \text{ and } \overrightarrow{c} = \hat{\mathrm{i}} - 2\hat{\mathrm{j}} + \hat{\mathrm{k}}$$
$$\therefore \overrightarrow{c} = (1, -2, 1)$$

Now, $\overrightarrow{a} + \overrightarrow{b} - 2\overrightarrow{c}$

$$= (2, 1, -1) + (1, -1, 2) - 2(1, -2, 1)$$
$$= (2, 1, -1) + (1, -1, 2) + (-2, 4, -2)$$
$$= (2 + 1 - 2, 1 - 1 + 4, -1 + 2 - 1)$$
$$\therefore \overrightarrow{a} + \overrightarrow{b} - 2\overrightarrow{c} = (1, 4, -1)$$
$$\therefore \left|\overrightarrow{a} + \overrightarrow{b} - 2\overrightarrow{c}\right| = |(1, 4, -1)|$$
$$= \sqrt{(1)^2 + (4)^2 + (-1)^2}$$

($\because$ Definition of magnitude)

$$= \sqrt{1 + 16 + 1}$$
$$= \sqrt{18}$$
$$= \sqrt{9(2)}$$
$$= 3\sqrt{2}$$

(ix) Here $\overrightarrow{a} = (1, 2, 1)$, $\overrightarrow{b} = (2, 1, 1)$, $\overrightarrow{c} = (3, 4, 1)$ are given

$$\therefore \overrightarrow{a} + 2\overrightarrow{b} + \overrightarrow{c}$$
$$= (1, 2, 1) + 2(2, 1, 1) + (3, 4, 1)$$
$$= (1, 2, 1) + (4, 2, 2) + (3, 4, 1)$$
$$= (1 + 4 + 3, 2 + 2 + 4, 1 + 2 + 1)$$
$$\therefore \overrightarrow{a} + 2\overrightarrow{b} - \overrightarrow{c} = (8, 8, 4)$$
$$\therefore \left|\overrightarrow{a} + 2\overrightarrow{b} + \overrightarrow{c}\right| = \sqrt{(8)^2 + (8)^2 + (4)^2}$$

($\because$ Definition of magnitude)

$$= \sqrt{64 + 64 + 16} = \sqrt{144} = \sqrt{(12)^2}$$

$$\therefore \left|\vec{a} + 2\vec{b} + \vec{c}\right| = 12$$

(x) Given that $\vec{a} = \hat{j} + \hat{k} - \hat{i} = -\hat{i} + \hat{j} + \hat{k}$

$\therefore \vec{a} = -\hat{i} + \hat{j} + \hat{k} = (-1, 1, 1)$ and $\vec{b} = 2\hat{\mathrm{i}} + \hat{\mathrm{j}} - 3\hat{\mathrm{k}} = (2, 1, -3)$

$\therefore 2\vec{a} + 3\vec{b}$

$= 2(-1, 1, 1) + 3(2, 1, -3)$

$= (-2 + 6, 2 + 3, 2 - 9)$

$\therefore 2\vec{a} + 3\vec{b} = (4, 5, -7)$

$\therefore \left|2\vec{a} + 3\vec{b}\right| = (4, 5, -7)$

$= \sqrt{(4)^2 + (5)^2 + (-7)^2}$

($\because$ Definition of magnitude)

$= \sqrt{16 + 25 + 49} = \sqrt{90} = \sqrt{9 \times 10}$

$\therefore \left|2\vec{a} + 3\vec{b}\right| = 3\sqrt{10}$

(xi) Given that $\vec{a} = (1, 2, 1)$, $\vec{b} = (1, -1, 2)$, $\vec{c} = (3, 2, -1)$then find

$\left|3\vec{a} + \vec{b} - 2\vec{c}\right|$

$\therefore 3\vec{a} + \vec{b} - 2\vec{c}$

$= 3(1, 2, 1) + (1, -1, 2) - 2(3, 2, -1)$

$= (3, 6, 3) + (1, -1, 2) - (6, 4, -2)$

$= (3 + 1 - 6, 6 - 1 - 4, 3 + 2 + 2)$

$= (-2, 1, 7)$

Hence, $\left|3\vec{a} + \vec{b} - 2\vec{c}\right| = \sqrt{4 + 1 + 49} = \sqrt{54}$

Illustration 1.20: If $a(1, 0, 0) + b(0, 1, 0) + c(2, -3, -7) = (0, 0, 0)$, where $a, b, c \in R$, then find the values of a, b, and c.

Solution: Here

$$a(1, 0, 0) + b(0, 1, 0) + c(2, -3, -7) = (0, 0, 0)$$

$$\therefore a\,(a,0,0) + b\,(0,b,0) + c\,(2c,-3c,-7c) = (0,0,0)$$

$$\therefore (a+0+2c, 0+b-3c, 0+0-7c) = (0,0,0)$$

$$\therefore (a+2c, b-3c, -7c) = (0,0,0)$$

$$\therefore a+2c = 0, b-3c = 0, -7c = 0$$

$\therefore c = 0, a = 0, b = 0$ or $a = 0, b = 0, c = 0$.

Illustration 1.21: A space shuttle of 1000 tons hangs from two skyscrapers using steel cables as shown in Figure 1.16. Find the forces or tensions in both the cables attached with skyscrapers and also find their magnitude.

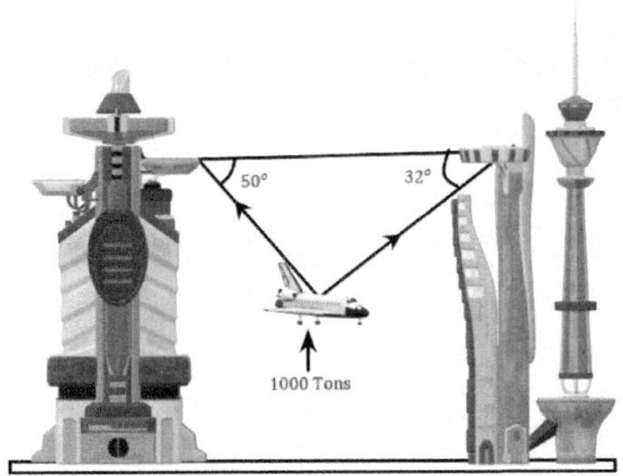

Figure 1.16 A space shuttle of 1000 tons weight hangs from two skyscrapers using steel cables

Solution:
Let $\overrightarrow{F_1}$ and $\overrightarrow{F_2}$ be two forces or tensions on the steel cables respectively. First, we represent $\overrightarrow{F_1}$ and $\overrightarrow{F_2}$ in terms of vertical and horizontal components.

$$\overrightarrow{F_1} = -\left|\overrightarrow{F_1}\right| \cos 50^o\,\hat{i} + \left|\overrightarrow{F_1}\right| \sin 50^o\,\hat{j}$$

$$\overrightarrow{F_2} = \left|\overrightarrow{F_2}\right| \cos 32^o\,\hat{i} + \left|\overrightarrow{F_2}\right| \sin 32^o\,\hat{j}$$

The gravity force acting on the space shuttle is $\overrightarrow{F} = -mg\hat{j} = -(1000)(9.8)\,\hat{j} = -9800\hat{j}$. Therefore, the counterbalance of $\overrightarrow{F}$ with $\overrightarrow{F_1}$ and $\overrightarrow{F_2}$ is given as

$$\therefore\ \overrightarrow{F_1} + \overrightarrow{F_2} + \overrightarrow{F} = 0 \Rightarrow \overrightarrow{F_1} + \overrightarrow{F_2} = -\overrightarrow{F} = -(-9800)\,\hat{j} = 9800\hat{j}$$

Thus,

$$\left(-\left|\overrightarrow{F_1}\right|\cos 50^o\,\hat{\imath}+\left|\overrightarrow{F_1}\right|\sin 50^o\,\hat{\jmath}\right)+$$
$$\left(\left|\overrightarrow{F_2}\right|\cos 32^o\,\hat{\imath}+\left|\overrightarrow{F_2}\right|\sin 32^o\,\hat{\jmath}\right)=9800\hat{\jmath}$$
$$\therefore\left(-\left|\overrightarrow{F_1}\right|\cos 50^o+\left|\overrightarrow{F_2}\right|\cos 32^o\right)\hat{\imath}+$$
$$\left(\left|\overrightarrow{F_1}\right|\sin 50^o+\left|\overrightarrow{F_2}\right|\sin 32^o\right)\hat{\jmath}=9800\hat{\jmath}$$

Now, equation the components

$$-\left|\overrightarrow{F_1}\right|\cos 50^o+\left|\overrightarrow{F_2}\right|\cos 32^o=0\Rightarrow\left|\overrightarrow{F_1}\right|\cos 50^o=\left|\overrightarrow{F_2}\right|\cos 32^o$$
$$\left|\overrightarrow{F_1}\right|\sin 50^o+\left|\overrightarrow{F_2}\right|\sin 32^o=9800$$

Solving for $|\overrightarrow{F_2}|$, we get

$$\left|\overrightarrow{F_1}\right|\sin 50^o+\frac{\left|\overrightarrow{F_1}\right|\cos 50^o}{\cos 32^o}\sin 32^o=9800$$
$$\therefore\left|\overrightarrow{F_1}\right|=\frac{9800}{\sin 50^o+\tan 32^o\cos 50^o}\approx 8392\ N$$

And

$$\left|\overrightarrow{F_2}\right|=\frac{\left|\overrightarrow{F_1}\right|\cos 50^o}{\cos 32^o}\approx 6361\ N$$

Thus, the force vectors are
$\overrightarrow{F_1}\approx -5394\,\hat{\imath}+6429\,\hat{\jmath}$ and $\overrightarrow{F_2}\approx 5394\hat{\imath}+3371\hat{\jmath}$.

1.14 Exercise

1. If $\overrightarrow{x}=(2,1)$ and $\overrightarrow{y}=(1,3)$, then (i) find a unit vector in the direction of $3\overrightarrow{x}-2\overrightarrow{y}$, (ii) find direction cosines of $3\overrightarrow{x}-2\overrightarrow{y}$.

$$\left(\text{Answer : (i) }\tfrac{4}{5}\hat{\imath}-\tfrac{3}{5}\hat{\jmath},\text{ (ii) }\tfrac{4}{5},-\tfrac{3}{5}\right)$$

2. If $x(3,2)+y(2,3)=(17,13)$, find the real values of x and y.

$$(\text{Answer : }(x,y)=(5,1))$$

3. Find $a, b \in R$ such that (i) $(4, 7) + (a, b) = (17, 13)$ (ii) $(a, -8) - 2(3, b) = (-4, 6)$.

   (Answer : (i) $(a, b) = (13, 20)$, (ii) $(a, b) = (2, -7)$)

4. If $\hat{x} = (4, 7, 2)$ and $\hat{y} = (-1, 3, 4)$, find the vectors $2\hat{x} + 4\hat{y}$ and $3\hat{x} - \hat{y}$.

   (Answer : (i) $2\overrightarrow{x} + 4\overrightarrow{y} = (4, 26, 20)$, (ii) $3\overrightarrow{x} - \overrightarrow{y} = (13, 18, 21)$)

5. If for real values of x, y, z, $x(1, 2, 0) + y(0, 3, 1) + z(-1, 0, 1) = (2, 1, 0)$, then find x, y, z.

   (Answer : $(x, y, z) = (5, -3, 3)$)

6. Find the vectors $\boldsymbol{x}, \boldsymbol{y} \in R^2$such that $|\boldsymbol{x}| = |\boldsymbol{y}| = 1$and $|\boldsymbol{x} - \boldsymbol{y}| = 2$.

   (Answer : $(1, 0)$ and $(-1, 0)$)

7. If $P(2, 4, -5)$, and $Q(1, 2, 3)$ are points in R^2, find the direction cosines of PQ.

   $\left(\text{Answer : } -\frac{1}{\sqrt{69}}, -\frac{2}{\sqrt{69}}, \frac{8}{\sqrt{69}}\right)$

8. If $\hat{a} = (3, -1, -4), \hat{b} = (-2, 4, -3), \hat{c} = (-1, 2, -1)$, then find the direction cosines of the vector $3\hat{a} - 2\hat{b} + 4\hat{c}$.

   $\left(\text{Answer : } \frac{9}{\sqrt{190}}, -\frac{3}{\sqrt{190}}, \frac{-10}{\sqrt{190}}\right)$

9. If $\hat{a} = (1, 2, 3), \hat{b} = (2, -2, -5), \hat{c} = (3, -2, -1)$, then find (i) $\hat{a} + 2\hat{b} - \hat{c}$ (ii) $\left|\hat{a} + \hat{b} + \hat{c}\right|$.

   (Answer : (i) $\sqrt{68}$ (ii) $\sqrt{41}$)

10. Show that $\hat{a} = (2, -3, 2), \hat{b} = \left(1, -\frac{3}{2}, 1\right)$ are parallel vectors

2

Scalar and Vector Products

2.1 Scalar Product, or Dot Product, or Inner Product

(1) **Algebraic definition:** If $\overrightarrow{x} = x_1\hat{i} + x_2\hat{j} + x_3\hat{k} = (x_1, x_2, x_3)$ and $\overrightarrow{y} = y_1\hat{i} + y_2\hat{j} + y_3\hat{k} = (y_1, y_2, y_3)$ are two vectors in R^3, then their dot product $\overrightarrow{x} \cdot \overrightarrow{y}$ (read $\overrightarrow{x}$ dot $\overrightarrow{y}$) is defined as

$$\overrightarrow{x} \cdot \overrightarrow{y} = x_1y_1 + x_2y_2 + x_3y_3 \tag{2.1}$$

Note that the result of a dot product is a scalar i.e., it is a real number.

(2) **Geometric definition:** If the angle between the vectors $\overrightarrow{a}$ and $\overrightarrow{b}$ is θ as shown in Figure 2.1 then

$$\overrightarrow{a}.\overrightarrow{b} = |\overrightarrow{a}|\left|\overrightarrow{b}\right|\cos\theta = ab\cos\theta, \tag{2.2}$$

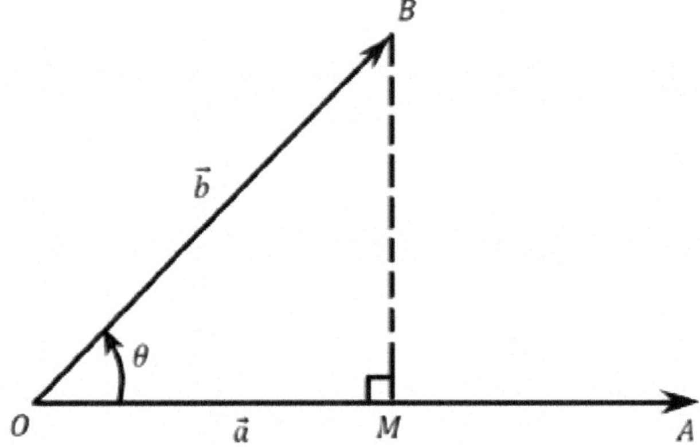

Figure 2.1 Represents a scalar or dot product

where, $|\vec{a}| = a$ and $\left|\vec{b}\right| = b$.

Note: The angle θ is denoted by $\left(\vec{a}\,\hat{;}\,\vec{b}\right)$ (read $\vec{a}$ cab $\vec{b}$) also.

2.2 The Measure of Angle Between two Vectors and Projections

Let $OA = \vec{a}, OB = \vec{b}$, and $\left(\vec{a}\,\hat{;}\,\vec{b}\right) = \theta$ as shown in Figure 2.1.

Now,

$$\vec{a} \cdot \vec{b} = \vec{a}\,\vec{b} \cos\theta = \vec{a}\,(OB \cos\theta) = \vec{a}\left(OB \times \frac{OM}{OB}\right) = \vec{a} \cdot OM$$

$\therefore \vec{a} \cdot \vec{b} = |\vec{a}|$ (Projection of $\vec{b}$ on the direction of $\vec{a}$)

Similarly, $\vec{a} \cdot \vec{b} = \left|\vec{b}\right|$ (Projection of $\vec{a}$ on the direction of $\vec{b}$)

Also, the projection $\vec{a}$ on the direction of $\vec{b} = \frac{\vec{a}\,.\,\vec{b}}{\left|\vec{b}\right|}$ and the projection of $\vec{b}$ on the direction of $\vec{a} = \frac{\vec{a}\,.\,\vec{b}}{|\vec{a}|}$.

Thus, geometrically, $\vec{a} \cdot \vec{b}$ denoted the product of the modulus of one vector and the projection of the second vector in the direction of the first vector.

An important result: From (2.2), we get $\cos\theta = \frac{\vec{a} \cdot \vec{b}}{|\vec{a}||\vec{b}|}$.

2.2.1 Properties of a Dot Product

(1) The scalar product of two vectors is commutative.

Thus, $\vec{a} \cdot \vec{b} = \vec{b} \cdot \vec{a}$

We have $\vec{a} \cdot \vec{b} = x_1y_1 + x_2y_2 + x_3y_3$

$$= y_1x_1 + y_2x_2 + y_3x_3$$

$$= \vec{b} \cdot \vec{a}$$

(2) If two non-zero vectors are perpendicular to each other then their dot product is zero.

If $\vec{a} \perp \vec{b}$, then $\theta = 90^\circ$ and $\cos\theta = 0$.

$$\therefore \vec{a} \cdot \vec{b} = ab \cos\theta = 0$$

Also, if $\vec{a} \cdot \vec{b} = 0$, then $ab \cos\theta = 0$.

i.e., $\cos\theta = 0 \Rightarrow \theta = \frac{\pi}{2}$.

(3) Scalar product of two like vectors or two opposite vectors:

(i) If $\vec{a}$ and $\vec{b}$ are of the same direction, then $\theta = 0$ and $\cos\theta = \cos\theta = 1$.

$$\therefore \vec{a} \cdot \vec{b} = ab\cos\theta = \vec{a}\,\vec{b} = \text{Product of their moduli}$$

(ii) If $\vec{a}$ and $\vec{b}$ have opposite directions, then $\theta = p$ and $\cos\theta = \cos\pi = -1$.

$$\therefore \vec{a} \cdot \vec{b} = ab\cos\theta = \vec{a}\,\vec{b}\,(-1)$$

$$= -\vec{a}\,\vec{b} = -(\text{Product of their moduli})$$

From (i), if $\vec{a} = \vec{b}$, then $\vec{a} \cdot \vec{a} = \vec{a}\,\vec{a} = \vec{a}^2 = |\vec{a}|^2$

(4) From unit vectors i, j, k in the directions of the axes:

$$\hat{i} \cdot \hat{i} = \hat{j} \cdot \hat{j} = \hat{k} \cdot \hat{k} = 1,$$

$$\hat{i} \cdot \hat{j} = 0, \hat{j} \cdot \hat{k} = 0, \hat{k} \cdot \hat{i} = 0,$$

Because $\hat{i}, \hat{j}, \hat{k}$ are mutually perpendicular.

(5) For $p, q \in R$,

$$p\vec{a} \cdot p\vec{b} = pq\left(\vec{a} \cdot \vec{b}\right) = (pq\vec{a}) \cdot \vec{b} = \vec{a} \cdot \left(pq\,\vec{b}\right)$$

(6) The scalar product of two vectors is distributive with respect to the vector addition.

$$\vec{a} \cdot \left(\vec{b} + \vec{c}\right) = \vec{a} \cdot \vec{b} + \vec{a} \cdot \vec{c}$$

Also, $\vec{a} \cdot \left(\vec{b} - \vec{c}\right) = \vec{a} \cdot \vec{b} - \vec{a} \cdot \vec{c}$

Note: If $\vec{a} \neq \theta$ and $\vec{a} \cdot \left(\vec{b} - \vec{c}\right) = 0$, then $\vec{a} \perp \left(\vec{b} - \vec{c}\right)$ or $\vec{b} - \vec{c} = \theta$.

(7) The scalar product of two vectors is given by the sum of the products of their corresponding elements.

If $\vec{a} = (a_1, a_2, a_3)$ and $\vec{b} = (b_1, b_2, b_3)$, then

$$\vec{a} \cdot \vec{b} = (a_1, a_2, a_3) \cdot (b_1, b_2, b_3)$$

$$= a_1 b_1 + a_2 b_2 + a_3 b_3 = \Sigma a_i b_i$$

Also $\vec{a} = |\vec{a}| = \sqrt{a_1^2 + a_2^2 + a_3^2} = \sqrt{\Sigma a_i^2}$

$$\vec{b} = \left|\vec{b}\right| = \sqrt{b_1^2 + b_2^2 + b_3^2} = \sqrt{\Sigma b_i^2}$$

Now, $\vec{a} \cdot \vec{b} = ab \cos\theta$ gives

$$\cos\,\theta = \frac{\vec{a}\,.\,\vec{b}}{|\vec{a}|\left|\vec{b}\right|}$$

$$= \frac{a_1 b_1 + a_2 b_2 + a_3 b_3}{\sqrt{a_1^2 + a_2^2 + a_3^2}\sqrt{b_1^2 + b_2^2 + b_3^2}}$$

$$= \frac{\Sigma a_i bi}{\sqrt{\Sigma a_i^2}\sqrt{\Sigma b_i^2}}$$

$$= \frac{\Sigma a_i bi}{\sqrt{\Sigma a_i^2 \times \Sigma b_i^2}}$$

$$\therefore \theta = \cos^{-1}\left\{\frac{\Sigma a_i b_i}{\sqrt{\Sigma a_i^2}\sqrt{\Sigma b_i^2}}\right\}$$

Note:

(1) If (l_1, m_1, n_1) and (l_2, m_2, n_2) are the direction cosines of vectors $\vec{a}$ and $\vec{b}$ respectively. Then

$$\cos\,\theta = l_1 l_2 + m_1 m_2 + n_1 n_2 = \Sigma l_1 l_2$$

(2)

$$\sin^2\theta = 1 - \cos^2\theta$$

$$= 1 - \frac{\left(\vec{a} \cdot \vec{b}\right)^2}{a^2 b^2} = \frac{a^2 b^2 - \left(\vec{a} \cdot \vec{b}\right)^2}{a^2 b^2}$$

$$\therefore \sin\theta = \frac{\sqrt{a^2b^2 - \left(\vec{a}\cdot\vec{b}\right)^2}}{|\vec{a}|\left|\vec{b}\right|} = \frac{\sqrt{|\vec{a}|^2\left|\vec{b}\right|^2 - \left(\vec{a}\cdot\vec{b}\right)^2}}{|\vec{a}|\left|\vec{b}\right|}$$

Remember the following two important results:

(1) $\vec{x}\cdot\vec{x} = |\vec{x}|^2 (2) \vec{x}\cdot\vec{y} = 0 \Leftrightarrow \vec{x}\perp\vec{y}$

Illustration 2.1: If $\vec{x} = (1, 2, 3)$ and $\vec{y} = (2, 3, 4)$,then find (i) $\vec{x}\cdot\vec{y}$ and (ii) $(\vec{x}\,\hat{,}\,\vec{y})$.

Solution:

(i) $$\vec{x}\cdot\vec{y} = (1,2,3)\cdot(2,3,4)$$
$$= (1)(2) + (2)(3) + (3)(4)$$
$$= 2 + 6 + 12 = 20$$

(ii) $|\vec{x}|^2 = 1 + 4 + 9 = 14$

$$\therefore |\vec{x}| = \sqrt{14}$$
$$|\vec{y}|^2 = 4 + 9 + 16 = 29$$
$$\therefore |\vec{y}| = \sqrt{29}$$

Now,

$$\cos\theta = \cos(\vec{x}\,,\hat{}\,\vec{y}) = \frac{\vec{x}\,.\,\vec{y}}{|\vec{x}||\vec{y}|} = \frac{20}{\sqrt{14}\sqrt{29}}$$
$$\therefore \theta = (\vec{x},\hat{}\,\vec{y}) = \cos^{-1}\left(\frac{20}{\sqrt{14}\sqrt{29}}\right)$$

Illustration 2.2: If $\vec{x} = \hat{i} + 3\hat{j} + 2\hat{k}$ and $\vec{y} = 4\hat{i} - 2\hat{j} + \hat{k}$, then find (i) $\vec{x}\cdot\vec{y}$ (ii) the angle between $\vec{x}$ and $\vec{y}$.

Solution:

(i) $$x\cdot y = (1,3,2)\cdot(4,-2,1)$$
$$= (1)(4) + (3)(-2) + (2)(1)$$
$$= 4 - 6 + 2 = 0$$

Hence, x and y are perpendicular.

(ii) As $x \cdot y = 0$, $x \perp y$.

$\therefore$The angle between x and y is 90° or $\frac{\pi}{2}$.

Illustration 2.3: If $\overline{x} = (1, -2, 2)$ and $\overline{y} = (0, 0, -1)$, then verify that

(i) $|\overline{x}.\overline{y}| \leq |\overline{x}||\overline{y}|$

(ii) $|\overline{x} + \overline{y}| \leq |\overline{x}| + |\overline{y}|$

(iii) $|\overline{x} + \overline{y}| \geq |\overline{x}| + |\overline{y}|$

Solution: Here, $\overline{x} = (1, -2, 2)$ and $\overline{y} = (0, 0, -1)$.
$\therefore |\overline{x}| = \sqrt{1 + 4 + 4} = \sqrt{9} = 3$ and $|\overline{y}| = \sqrt{0 + 0 + 1} = 1$

(i)

$$\begin{aligned} \overline{x} \cdot \overline{y} &= (1, -2, 2) \cdot (0, 0, -1) \\ &= (1)\,(0) + (-2)\,(0) + (2)(-1) \\ &= -2 \end{aligned}$$

$$\therefore |\overline{x} \cdot \overline{y}| = |-2| = 2 \qquad (2.3)$$

Also, $|\overline{x}| = 3$ and $|\overline{y}| = 1$

$$\therefore |\overline{x}| \cdot |\overline{y}| = 3 \qquad (2.4)$$

Hence, from (2.3) and (2.4) we get $2 < 3 \Rightarrow |\overline{x} \cdot \overline{y}| \leq |\overline{x}||\overline{y}|$.

(ii) $\overline{x} + \overline{y} = (1, -2, 2) + (0, 0, -1)$

$$\begin{aligned} &= (1 + 0, -2 + 0, 2 + (-1)) \\ &= (1, -2, 1) \end{aligned}$$

$$\therefore |\overline{x} + \overline{y}| = \sqrt{1 + 4 + 1} = \sqrt{6} \qquad (2.5)$$

Also, $|\overline{x}| = 3$ and $|\overline{y}| = 1$

$$\therefore |\overline{x}| + |\overline{y}| = 3 + 1 = 4 \qquad (2.6)$$

Hence, from (2.5) and (2.6) we get $\sqrt{6} < 4 \;\; (\because 6 < 16)$

$$\therefore |\overline{x} + \overline{y}| \leq |\overline{x}| + |\overline{y}|.$$

(iii) $\overline{x} - \overline{y} = (1, -2, 2) - (0, 0, -1)$

$$= (1 - 0, -2 - 0, 2 - (-1))$$
$$= (1, -2, 3)$$
$$\therefore |\overline{x} - \overline{y}| = \sqrt{1 + 4 + 9} = \sqrt{14} \quad (2.7)$$

Also, $|\overline{x}| = 3$ and $|\overline{y}| = 1$

$$\therefore |\overline{x}| - |\overline{y}| = 3 - 1 = 2 \quad (2.8)$$

Hence, from (2.7) and (2.8) we get $\sqrt{14} > 2$ $(\because 14 > 4)$
Therefore, $|\overline{x} - \overline{y}| \geq |\overline{x}| - |\overline{y}|$.

Illustration 2.4: If for $\overrightarrow{a}, \overrightarrow{b} \in R^3$, $|\overrightarrow{a}| = |\overrightarrow{b}|$, then show that $(\overrightarrow{a} + \overrightarrow{b})$ and $(\overrightarrow{a} - \overrightarrow{b})$ are perpendicular to each other.

Solution: To prove $(\overrightarrow{a} + \overrightarrow{b})$ and $(\overrightarrow{a} - \overrightarrow{b})$ are perpendicular to each other, we have to show that $\left(\overrightarrow{a} + \overrightarrow{b}\right) \cdot \left(\overrightarrow{a} - \overrightarrow{b}\right) = 0$.

$$\therefore \left(\overrightarrow{a} + \overrightarrow{b}\right) \cdot \left(\overrightarrow{a} - \overrightarrow{b}\right) = \overrightarrow{a} \cdot \overrightarrow{a} - \overrightarrow{a} \cdot \overrightarrow{b} + \overrightarrow{b} \cdot \overrightarrow{a} + \overrightarrow{b} \cdot \overrightarrow{b}$$
$$= |\overrightarrow{a}|^2 - \overrightarrow{a} \cdot \overrightarrow{b} + \overrightarrow{b} \cdot \overrightarrow{a} + \left|\overrightarrow{b}\right|^2$$
$$(\because \overrightarrow{a} \cdot \overrightarrow{a} = |\overrightarrow{a}|^2 \; and \; \overrightarrow{b} \cdot \overrightarrow{b} = \left|\overrightarrow{b}\right|^2)$$
$$= |\overrightarrow{a}|^2 - \left|\overrightarrow{b}\right|^2 \qquad (\because |\overrightarrow{a}| = |\overrightarrow{b}|)$$
$$= 0$$
$$\therefore \left(\overrightarrow{a} + \overrightarrow{b}\right) \perp \left(\overrightarrow{a} - \overrightarrow{b}\right)$$

Illustration 2.5: If for $\overrightarrow{a} = (2, 2, -1)$ and $\overrightarrow{b} = (6, -3, 2)$, then find the angle between $\overrightarrow{a}$ and $\overrightarrow{b}$.

Solution: We have $\overrightarrow{a} = (2, 2, -1)$ and $\overrightarrow{b} = (6, -3, 2)$, then

$$\overrightarrow{a} . \overrightarrow{b} = (2, 2, -1) \cdot (6, -3, 2) = 12 - 6 - 2 = 4$$

Also, $|\overrightarrow{a}| = \sqrt{4 + 4 + 1} = \sqrt{9} = 3$, $\left|\overrightarrow{b}\right| = \sqrt{36 + 9 + 4} = \sqrt{49} = 7$,

If θ is the angle between $\vec{a}$ and $\vec{b}$, then

$$\cos\theta = \frac{\vec{a}\cdot\vec{b}}{|\vec{a}||\vec{b}|} = \frac{4}{3\times 7} = \frac{4}{21} = 0.1905$$

$$\therefore \sin\left(90^0 - \theta\right) = 0.1905$$

$$\therefore\ 90^0 - \theta = 10^0 58' \ (\because \text{Using trigonometric table})$$

$$\therefore \theta = 90^0 - 10^0 58' = 79^0 2'$$

Illustration 2.6: If $\vec{a} = 3\hat{i} - 2\hat{j} + \hat{k}$ and $\vec{b} = \hat{i} - 2\hat{j} + \hat{k}$, then find the projection of $\vec{a}$ on $\vec{b}$.

Solution: The projection of $\vec{a}$ on $\vec{b}$ is given as

$$= \frac{\vec{a}\cdot\vec{b}}{\left|\vec{b}\right|}$$

$$= \frac{(3,-2,1)\cdot(1,-2,1)}{\sqrt{1+4+1}}$$

$$= \frac{3+4+1}{\sqrt{6}} = \frac{8}{\sqrt{6}} = \frac{4\sqrt{6}}{3}$$

Illustration 2.7: Show that $(-1,6,6)$, $(-4,9,6)$, and $(0,7,10)$ are position vectors of the vertices of a right-angled triangle.

Solution: Let $A(-1,6,6)$, $B(-4,9,6)$, and C$(0,7,10)$ be the vertices of $\Delta\ ABC$.

Now, $AB = OB - OA = (-4,9,6) - (-1,6,6) = (-3,3,0)$

$$BC = OC - OB = (0,7,10) - (-4,9,6) = (4,-2,4)$$

And $CA = OA - OC = (-1,6,6) - (0,7,10) = (-1,-1,-4)$

$$AB\cdot CA = (-3,3,0)\cdot(-1,-1,-4) = 3 - 3 + 0 = 0$$

$$\therefore AB \perp CA$$

$\therefore \Delta\ ABC$ is a right-angled triangle.

Illustration 2.7: Find the angle between the vectors $2\hat{i}+\hat{j}+4\hat{k}$ and $\hat{i}+\hat{j}+\hat{k}$.

Solution: Let $\overrightarrow{x} = 2\hat{i} + \hat{j} + 4\hat{k} = (2, 1, 4)$ and $\overrightarrow{y} = \hat{i} + \hat{j} + \hat{k} = (1, 1, 1)$ then we have

$\overrightarrow{x} \cdot \overrightarrow{y} = (2, 1, 4) \cdot (1, 1, 1) = 2 + 1 + 4 = 7$ and if $(x, \hat{} \; y) = \theta$, then

$$\therefore \cos\theta = \frac{\overrightarrow{x} \cdot \overrightarrow{y}}{|\overrightarrow{x}||\overrightarrow{y}|} = \frac{7}{\sqrt{21}\,\sqrt{3}} = \frac{7}{\sqrt{63}} = \frac{\sqrt{7}}{3}$$

$$\therefore \theta = \cos^{-1}\frac{\sqrt{7}}{3}$$

Illustration 2.8: For what value of x, the vectors $2\hat{i} - 3\hat{j} + 5\hat{k}$ and $x\hat{i} - 6\hat{j} - 8\hat{k}$ are perpendicular to each other?

Solution: Let $\overrightarrow{a} = 2\hat{i} - 3\hat{j} + 5\hat{k} = (2, -3, 5)$ and $\overrightarrow{b} = x\hat{i} - 6\hat{j} - 8\hat{k} = (x, -6, -8)$.

As $\overrightarrow{a} \perp \overrightarrow{b} \Rightarrow \overrightarrow{a} \cdot \overrightarrow{b} = 0$

$$\therefore \; (2, -3, 5) \cdot (x, -6, -8) = 0$$

$$\therefore 2x + 18 - 40 = 0$$

$$\therefore 2x = 22 \Rightarrow x = 11.$$

2.3 Vector Product or Cross Product or Outer Product of Two Vectors

The vector product of two vectors $\overrightarrow{a}$ and $\overrightarrow{b}$ (See Figure 2.2) is denoted by $\overrightarrow{a} \times \overrightarrow{b}$ and it is defined as follow:

(1) **Algebraic definition**: If $\overrightarrow{a} = (a_1, a_2, a_3)$ and $\overrightarrow{b} = (b_1, b_2, b_3)$ then

$$\overrightarrow{a} \times \overrightarrow{b} = \begin{bmatrix} \hat{i} & \hat{j} & \hat{k} \\ a_1 & a_2 & a_3 \\ b_1 & b_2 & b_3 \end{bmatrix} \tag{2.9}$$

$$\overrightarrow{a} \times \overrightarrow{b} = (a_2b_3 - a_3b_2)\,\hat{i} - (a_1b_3 - a_3b_1)\,\hat{j} + (a_1b_2 - a_2b_1)\,\hat{k}$$

$$= (a_2b_3 - a_3b_2, a_3b_1 - a_1b_3, a_1b_2 - a_2b_1)$$

Note: $\overrightarrow{a} \times \overrightarrow{b}$ is also a vector.

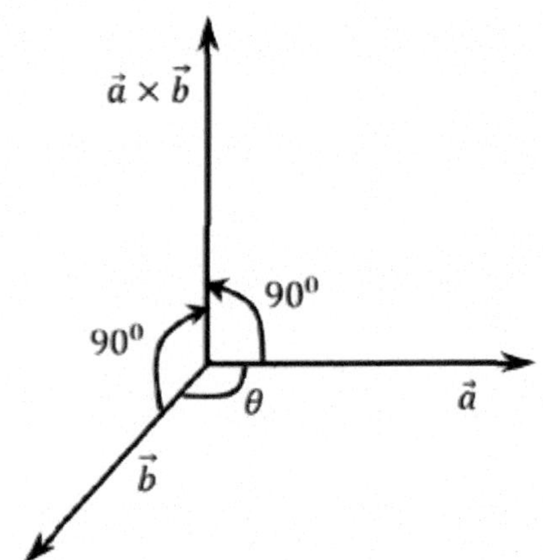

Figure 2.2 Represents a vector or cross product

(2) **Geometric definition:**

If $\vec{a}$ and $\vec{b}$ are given vectors, then their cross product $\vec{a} \times \vec{b}$ is defined as

$$\vec{a} \times \vec{b} = |\vec{a}| \left|\vec{b}\right| \sin\theta\, \hat{n} \tag{2.10}$$

where θ is an angle between $\vec{a}$ and $\vec{b}$, $\hat{n}$ is a unit vector perpendicular to both $\vec{a}$ and $\vec{b}$.

$$\therefore \vec{a} \times \vec{b} = (ab \sin\theta\,)\,\hat{n} \tag{2.11}$$

2.4 Geometric Interpretation of a Vector Product

The vector product of two vectors $\vec{a} \times \vec{b}$ has modulus $\left|\vec{a} \times \vec{b}\right|$ which is twice the area of the triangle whose two consecutive sides are $\vec{a}$ and $\vec{b}$ or equals the area of a parallelogram whose two consecutive sides are $\vec{a}$ and $\vec{b}$.

Note that $\vec{a} \times \vec{b}$ is a vector but $|\vec{a} \times \vec{b}|$ is a scalar that gives the area of the parallelogram $ABCD$ as shown in text is present for Figure 2.3.

$$\text{Thus, } \left|\vec{a} \times \vec{b}\right| = |ab \sin\theta \;\; \hat{n}|$$
$$= ab \sin\theta \quad (\because |\hat{n}| = 1)$$

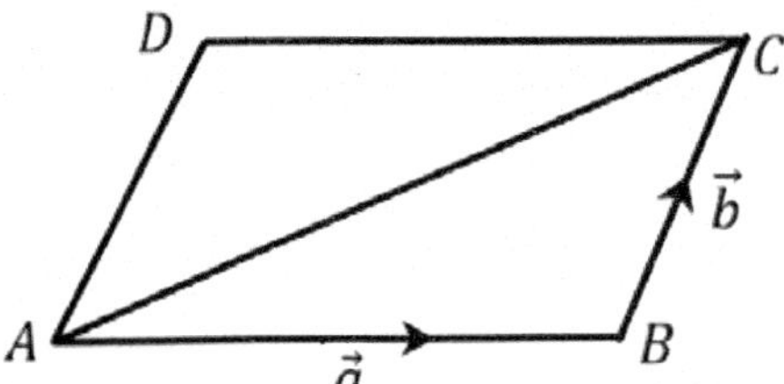

Figure 2.3 Represents the geometric interpretation of a vector or cross product

$$= \text{Area of the parallelogram } ABCD$$
$$= 2 \times \text{Area of } \triangle\ ABC$$

2.4.1 Properties of a Vector Product

1. Vector product or cross product is non-commutative.

   i.e., $\overrightarrow{a} \times \overrightarrow{b} \neq \overrightarrow{b} \times \overrightarrow{a}$.

   Here if θ is the angle between directions of $\overrightarrow{a}$ and $\overrightarrow{b}$, then the angle between the direction of $\overrightarrow{b}$ and $\overrightarrow{a}$ is $-\theta$.

   $$\therefore\ \overrightarrow{a} \times \overrightarrow{b} = ab \sin \theta\ .\ \hat{n}$$

   and $\overrightarrow{b} \times \overrightarrow{a} = ab \sin(-\theta)\ .\hat{n}\ = -ab \sin \theta\ .\ \hat{n}$

   $$(\because \sin(-\theta)\ = -\sin \theta\ )$$
   $$= -(\overrightarrow{a} \times \overrightarrow{b})$$

2. The vector product of two parallel vectors is a null vector.

   If $\overrightarrow{a} \parallel \overrightarrow{b}$, then $\theta = 0$.

   $$\therefore\ \overrightarrow{a} \times \overrightarrow{b} = ab \sin \theta\ \cdot \hat{n} = ab \sin 0\ \cdot\ \hat{n} = \theta \qquad (\because \sin 0\ = 0)$$

3. For the unit vectors $\hat{i},\ \hat{j},\ \hat{k}$:

   $$\hat{i} \times \hat{i} = \hat{j} \times \hat{j} = \hat{k} \times \hat{k} = \theta$$

   and $\hat{i} \times \hat{j} = \hat{k},\ \hat{j} \times \hat{k} = \hat{i},\ \hat{k} \times \hat{i} = \hat{j},\ \hat{j} \times \hat{i} = -\hat{k},\ \hat{k} \times \hat{j} = -\hat{i},$

   $$\hat{i} \times \hat{k} = -\hat{j}.$$

4. For any $p, q \in R$,

$$p\vec{a} \times q\vec{b} = pq\left(\vec{a} \times \vec{b}\right) = (pq\,\vec{a}) \times \vec{b} = \vec{a} \times (pq\,\vec{b})$$

Taking $\vec{a} = 1$ and $\vec{b} = -1$, we get

$$\vec{a} \times \left(-\vec{b}\right) = (-\vec{a}) \times \vec{b} = -(\vec{a} \times \vec{b})$$

5. Relation between cross product and dot product:

$$\left(\vec{a} \cdot \vec{b}\right)^2 = |\vec{a}|^2 \left|\vec{b}\right|^2 - \left|\vec{a} \times \vec{b}\right|^2$$

Or

$$\left(\vec{a} \cdot \vec{b}\right)^2 + \left|\vec{a} \times \vec{b}\right|^2 = |\vec{a}|^2\left|\vec{b}\right|^2$$

6. Distributive law for vector product:

(i) $\vec{a} \times \left(\vec{b} + \vec{c}\right) = \vec{a} \times \vec{b} + \vec{a} \times \vec{c}$

(ii) $\left(\vec{a} + \vec{b}\right) \times \vec{c} = \vec{a} \times \vec{c} + \vec{b} \times \vec{c}$

7. Expression of a cross product in terms of unit vectors $\hat{i}, \hat{j}, \hat{k}$:

If $\vec{a} = a_1\hat{i} + a_2\hat{j} + a_3\,\hat{k} = (a_1, a_2, a_3)$ and $\vec{b} = b_1\hat{i} + b_2\,\hat{j} + b_3\hat{k}$, then

$$\vec{a} \times \vec{b} = \begin{vmatrix} \hat{i} & \hat{j} & \hat{k} \\ a_1 & a_2 & a_3 \\ b_1 & b_2 & b_3 \end{vmatrix}$$

$$= (a_2b_3 - a_3b_2)\,\hat{i} - (a_1b_3 - a_3b_1)\,\hat{j} + (a_1b_2 - a_2b_1)\,\hat{k}$$

Remarks:

1. As $\left|\vec{a} \times \vec{b}\right| = ab\sin\theta$, $\sin\theta = \frac{|\vec{a} \times \vec{b}|}{|\vec{a}||\vec{b}|}$

2. If the direction cosines of $\vec{a}$ and $\vec{b}$ are (l_1, m_1, n_1) and (l_2, m_2, n_2) respectively, then

$$\sin\theta = \sqrt{(m_1n_2 - m_2n_1)^2 + (l_1n_2 - l_2n_1)^2 + (l_1m_2 - l_2m_1)^2}$$

3. By definition of the cross product,

$\vec{a} \perp (\vec{a} \times \vec{b})$ and $\vec{b} \perp (\vec{a} \times \vec{b})$

$\therefore \vec{a} \cdot \left(\vec{a} \times \vec{b}\right) = 0$ and $\vec{b} \cdot \left(\vec{a} \times \vec{b}\right) = 0.$

Illustration 2.9: Find the modulus of $\left(2\hat{i} - 3\hat{j} + \hat{k}\right) \times \left(\hat{i} - \hat{j} + 2\hat{k}\right)$.

Solution: Consider $\vec{a} = \left(2\hat{i} - 3\hat{j} + \hat{k}\right)$ and $\vec{b} = \left(\hat{i} - \hat{j} + 2\hat{k}\right)$

$$\therefore \vec{a} \times \vec{b} = \begin{vmatrix} \hat{i} & \hat{j} & \hat{k} \\ 2 & -3 & 1 \\ 1 & -1 & 2 \end{vmatrix}$$

$$= (((-3) \times 2) - ((-1) \times 1))\hat{i} - ((2 \times 2) - (1 \times 1))\hat{j}$$

$$+ ((2 \times (-1)) - (1 \times (-3)))\hat{k}$$

$$= (-6 + 1)\,\hat{i} - 3\hat{j} + \hat{k} = -5\hat{i} - 3\hat{j} + \hat{k}$$

$$\therefore \left|\vec{a} \times \vec{b}\right| = \sqrt{25 + 9 + 1} = \sqrt{35}$$

Illustration 2.10: If $\vec{a} = 2\hat{i} - \hat{j},\ \vec{b} = \hat{i} + 3\hat{j} - 2\hat{k}$ then compute

$$\left|\left(\vec{a} + \vec{b}\right) \times \left(\vec{a} - \vec{b}\right)\right|.$$

Solution: Here, $\vec{a} = 2\hat{i} - \hat{j} = (2, -1, 0)$ and $\vec{b} = \hat{i} + 3\hat{j} - 2\hat{k} = (1, 3, -2)$.

$$\therefore\ \vec{a} + \vec{b} = (2, -1, 0) + (1, 3, -2) = (3, 2, -2) \text{ and}$$

$$\vec{a} - \vec{b} = (2, -1, 0) - (1, 3, -2) = (1, -4, 2)$$

$$\therefore \left(\vec{a} + \vec{b}\right) \times \left(\vec{a} - \vec{b}\right) = \begin{vmatrix} \hat{i} & \hat{j} & \hat{k} \\ 3 & 2 & -2 \\ 1 & -4 & 2 \end{vmatrix}$$

$$= \hat{i}\,(4 - 8) - \hat{j}\,(6 + 2) + \hat{k}(-12 - 2)$$

$$= -4\hat{i} - 8\hat{j} - 14\hat{k} = (-4, -8, -14)$$

$$\therefore \left|\left(\vec{a} + \vec{b}\right) \times \left(\vec{a} - \vec{b}\right)\right| = \sqrt{16 + 64 + 196} = \sqrt{276} = 2\sqrt{69}$$

Illustration 2.11: Simplify:

$$\left(10\hat{i}+2\hat{j}+3\hat{k}\right)\cdot\left[\left(\hat{i}-2\hat{j}+2\hat{k}\right)\times(3\hat{i}-2\hat{j}-2\hat{k})\right].$$

Solution: Consider $\vec{a} = 10\hat{i}+2\hat{j}+3\hat{k} = (10,2,3)$, $\vec{b} = \hat{i}-2\hat{j}+2\hat{k} = (1,-2,2)$ and $\vec{c} = 3\hat{i}-2\hat{j}-2\hat{k} = (3,-2,-2)$.

$$\therefore \vec{b}\times\vec{c} = \begin{vmatrix} \hat{i} & \hat{j} & \hat{k} \\ 1 & -2 & 2 \\ 3 & -2 & -2 \end{vmatrix} = 8\hat{i}+8\hat{j}+4\hat{k} = 4(2,2,1)$$

The given expression

$$\vec{a}\cdot\left(\vec{b}\times\vec{c}\right) = (10,2,3)\cdot 4\,(2,2,1) = 4\,(20+4+3) = 108.$$

Illustration 2.12: Prove that

$$\vec{a}\times\left(\vec{b}+\vec{c}\right)+\vec{b}\times(\vec{c}+\vec{a})+\vec{c}\times\left(\vec{a}+\vec{b}\right) = 0.$$

Solution: $L.H.S. = \vec{a}\times\left(\vec{b}+\vec{c}\right)+\vec{b}\times(\vec{c}+\vec{a})+\vec{c}\times\left(\vec{a}+\vec{b}\right)$

Using distributive law

$$= \vec{a}\times\vec{b}+\vec{a}\times\vec{c}+\vec{b}\times\vec{c}+\vec{b}\times\vec{a}+\vec{c}\times\vec{a}+\vec{c}\times\vec{b}$$

$$= \vec{a}\times\vec{b}-\vec{c}\times\vec{a}-\vec{c}\times\vec{b}-\vec{a}\times\vec{b}+\vec{c}\times\vec{a}+\vec{c}\times\vec{b} = 0 = R.H.S.$$

Illustration 2.13: Show that: $(\vec{x}-\vec{y})\times(\vec{x}+\vec{y}) = 2(\vec{x}\times\vec{y})$

Solution: L.H.S. $= (\vec{x}-\vec{y})\times(\vec{x}+\vec{y})$

$$= \vec{x}\times(\vec{x}+\vec{y})-\vec{y}\times(\vec{x}+\vec{y})$$

$$= \vec{x}\times\vec{x}+\vec{x}\times\vec{y}-\vec{y}\times\vec{x}-\vec{y}\times\vec{y}$$

Now, we use $\vec{x}\times\vec{x} = \theta$, $\vec{y}\times\vec{y} = \theta$ and $-\vec{y}\times\vec{x} = \vec{x}\times\vec{y}$ in the above expression

$$= \theta+\vec{x}\times\vec{y}+\vec{x}\times\vec{y}-\theta = 2\,(\vec{x}\times\vec{y}) = R.H.S$$

Illustration 2.14: If $\vec{x} = (3,-1,2)$ and $\vec{y} = (2,1,-1)$ are given vectors. Find unit perpendicular vector to the $\vec{x}$ and $\vec{y}$ both.

Solution: Given that $\overrightarrow{x} = (3, -1, 2)$ and $\overrightarrow{y} = (2, 1, -1)$

$$\therefore \overrightarrow{x} \times \overrightarrow{y} = \begin{vmatrix} \hat{\imath} & \hat{\jmath} & \hat{k} \\ 3 & -1 & 2 \\ 2 & 1 & -1 \end{vmatrix} = (1-2)\,\hat{\imath} - (-3-4)\,\hat{\jmath} + (3+2)\,\hat{k}$$

$$= -\hat{\imath} + 7\hat{\jmath} + 5\hat{k} = (-1, 7, 5)$$

$$|\overrightarrow{x} \times \overrightarrow{y}| = \sqrt{1+49+25} = \sqrt{75} = 5\sqrt{3}$$

$\therefore$ Unit perpendicular vector to given vectors is $\overrightarrow{x}$ and $\overrightarrow{y}$

$$= \frac{\overrightarrow{x} \times \overrightarrow{y}}{|\overrightarrow{x} \times \overrightarrow{y}|} = \frac{1}{5\sqrt{3}}(-1, 7, 5).$$

Illustration 2.15: Find a unit vector of magnitude 10 which is perpendicular to vectors $2\hat{\imath} - \hat{\jmath} - 2\hat{k}$ and $4\hat{\imath} - 3\hat{\jmath} - 5\hat{k}$.

Solution: Consider the vectors $\overrightarrow{a} = 2\hat{\imath} - \hat{\jmath} - 2\hat{k} = (2, -1, -2)$ and $\overrightarrow{b} = 4\hat{\imath} - 3\hat{\jmath} - 5\hat{k} = (4, -3, -5)$

$$\text{Now, } \overrightarrow{a} \times \overrightarrow{b} = \begin{vmatrix} \hat{\imath} & \hat{\jmath} & \hat{k} \\ 2 & -1 & -2 \\ 4 & -3 & -5 \end{vmatrix}$$

$$= (5-6)\,\hat{\imath} - (-10+8)\,\hat{\jmath} + (-6+4)\,\hat{k}$$

$$= -\hat{\imath} + 2\hat{\jmath} - 2\hat{k} = (-1, 2, -2)$$

$$\therefore \left|\overrightarrow{a} \times \overrightarrow{b}\right| = \sqrt{1+4+4} = \sqrt{9} = 3$$

Now, the unit perpendicular vector to $\overrightarrow{a}$ and $\overrightarrow{b}$

$$= \frac{\overrightarrow{a} \times \overrightarrow{b}}{|\overrightarrow{a} \times \overrightarrow{b}|} = \frac{1}{3}(-1, 2, -2)$$

A perpendicular vector with magnitude 10 to both $\overrightarrow{a}$ and $\overrightarrow{b}$

$$= \frac{10}{3}(-1, 2, -2).$$

Illustration 2.16: If $\overrightarrow{x} = \hat{\imath} + \hat{\jmath} + \hat{k}$ and $\overrightarrow{y} = 2\hat{\imath} - \hat{\jmath} - \hat{k}$, then prove that $\overrightarrow{x}$ and $\overrightarrow{y}$ are perpendicular to each other. Also, find a unit vector perpendicular to the vectors $\overrightarrow{x}$ and $\overrightarrow{y}$.

Solution: Given that

$\vec{x} = \hat{i} + \hat{j} + \hat{k} = (1,1,1)$ and $\vec{y} = 2\hat{i} - \hat{j} - \hat{k} = (2,-1,-1)$

Now, $\vec{x}\,.\,\vec{y} = (1,1,1)\,.\,(2,-1,-1) = 2 - 1 - 1 = 0 \Rightarrow \vec{x} \perp \vec{y}$.

Also, $\vec{x} \times \vec{y} = \begin{vmatrix} \hat{i} & \hat{j} & \hat{k} \\ 1 & 1 & 1 \\ 2 & -1 & -1 \end{vmatrix}$

$$= (-1+1)\,\hat{i} - (-1-2)\,\hat{j} + (-1-2)\,\hat{k}$$

$$= 3\hat{j} - 3\hat{k} = (0,3,-3)$$

$$|\vec{x} \times \vec{y}| = \sqrt{0+9+9} = \sqrt{18} = 3\sqrt{2}$$

Unit perpendicular vector to $\vec{x}$ and $\vec{y}$

$$= \frac{\vec{x} \times \vec{y}}{|\vec{x} \times \vec{y}|} = \frac{1}{3\sqrt{2}}\left(3\hat{j} - 3\hat{k}\right) = \frac{1}{\sqrt{2}}\left(\hat{j} - \hat{k}\right).$$

Illustration 2.17: Find a unit vector that makes an angle of 60^0 with the vector $\hat{i} - \hat{k}$.

Solution: Let $\vec{x} = \hat{i} - \hat{k}$.

$$\therefore \vec{x} = (1,0,-1)$$

And suppose $\vec{y} = (y_1, y_2,\ y_3)$ be the required vector.

$$\therefore |\vec{y}|^2 = y_1^2 + y_2^2 + y_3^2 = 1 \tag{2.12}$$

Now, $(\vec{x}\,\hat{;}\,\vec{y}) = 60^0$

$$\cos 60^0 = \cos\left(\vec{x}\,\hat{;}\,\vec{y}\right) = \frac{\vec{x}\cdot\vec{y}}{|\vec{x}||\vec{y}|}$$

$$= \frac{(1,0,-1).\,\vec{y}}{\sqrt{2}\cdot 1}$$

$$\therefore \frac{1}{2} = \frac{y_1 - y_3}{\sqrt{2}}$$

$$\therefore y_1 - y_3 = \frac{1}{\sqrt{2}}$$

If we take $y_1 = 0$, then $y_3 = -\frac{1}{\sqrt{2}}$. Then from (2.12), we get $y_2^2 = 1 - \frac{1}{2} = \frac{1}{2} \Rightarrow y_2 = \frac{1}{\sqrt{2}}$. Thus, the required vector $\overrightarrow{y} = (0, -\frac{1}{\sqrt{2}}, \frac{1}{\sqrt{2}})$.

Illustration 2.18: Show that the angle between the vectors $\hat{\imath}+2\hat{\jmath}$ and $\hat{\imath}+\hat{\jmath}+3\hat{k}$ is $\sin^{-1}\sqrt{\frac{46}{55}}$.

Solution: Consider the vectors $\overrightarrow{a} = \hat{\imath}+2\hat{\jmath} = (1,2,0)$ and $\overrightarrow{b} = \hat{\imath}+\hat{\jmath}+3\hat{k} = (1,1,3)$.

Also, $|\overrightarrow{a}| = \sqrt{1+4+0} = \sqrt{5}$, $\left|\overrightarrow{b}\right| = \sqrt{1+1+9} = \sqrt{11}$

$$\overrightarrow{a} \times \overrightarrow{b} = \begin{vmatrix} \hat{\imath} & \hat{\jmath} & \hat{k} \\ 1 & 2 & 0 \\ 1 & 1 & 3 \end{vmatrix} = (6-0)\,\hat{\imath} - (3-0)\,\hat{\jmath} + (1-2)\,\hat{k}$$

$$= 6\hat{\imath} - 3\hat{\jmath} - \hat{k} = (6, -3, -1)$$

$$\therefore \left|\overrightarrow{a} \times \overrightarrow{b}\right| = \sqrt{36+9+1} = \sqrt{46}$$

Now, if $\left(\overrightarrow{a} \,\widehat{,}\, \overrightarrow{b}\right) = \theta$, then

$$\sin\theta = \frac{|\overrightarrow{a} \times \overrightarrow{b}|}{|\overrightarrow{a}||\overrightarrow{b}|} = \frac{\sqrt{46}}{\sqrt{5}\,\sqrt{11}} = \sqrt{\frac{46}{55}}$$

$$\therefore \theta = \sin^{-1}\sqrt{\frac{46}{55}}$$

2.5 Application of Scalar and Vector Products

The scalar and vector products are useful in many fields such as coordinate geometry, solid geometry, trigonometry, mechanics, physics, computer science, electric engineering, etc. In this section, we have discussed two basic applications of scalar and vector products

(1) Work done by a force

(2) Moment of a force about a point.

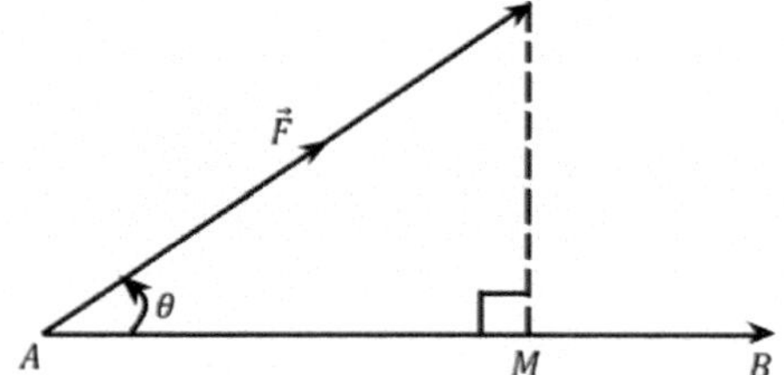

Figure 2.4 Represents work done by a force $\vec{F}$ on a particle

2.5.1 Work Done by a Force

The scalar product of a force applied to a particle and the displacement of the particle in the direction of the force is known as work. Thus, work is a scalar quantity.

If $\overrightarrow{F}$ denotes the force applied to a particle and $\overrightarrow{d}$ denotes the displacement of the particle in the direction of $\overrightarrow{F}$, then work done (See Figure 2.4) by the force $\overrightarrow{F}$ is given by

$$W = \overrightarrow{F} \cdot \overrightarrow{d}$$

Suppose a force $\overrightarrow{F} = (F_1, F_2, F_3)$ acts on a particle at the point $A(a_1, a_2, a_3)$ and as a result, the particle moves to the point $B\,(b_1, b_2, b_3)$.

$\therefore$ The displacement $\overrightarrow{d} = AB = OB - OA = (b_1 - a_1, b_2 - a_2, b_3 - a_3)$

Now, work done by the force $\overrightarrow{F}$ is

$$\begin{aligned} W &= \left(\text{Projection of } \overrightarrow{F} \text{ on } AB\right) \times AB \\ &= AM \times AB = \frac{\overrightarrow{F} \cdot \overrightarrow{AB}}{AB} \times AB \\ &= \overrightarrow{F} \cdot \overrightarrow{AB} \\ &= \text{Dot product of } \overrightarrow{F} \text{ and } \overrightarrow{AB} \\ &= (F_1, F_2, F_3) \cdot \; (b_1 - a_1, b_2 - a_2, b_3 - a_3) \\ &= F_1\,(b_1 - a_1) + F_2\,(b_2 - a_2) + F_3(b_3 - a_3) \end{aligned}$$

Note: If forces $\overrightarrow{F_1}, \overrightarrow{F_2}, \overrightarrow{F_3}, \ldots$ act on the point A, then $\overrightarrow{F} = \overrightarrow{F_1} + \overrightarrow{F_2} + \overrightarrow{F_3} + \ldots$

2.5.2 Moment of a Force About a Point

The moment of a force $\overrightarrow{F}$ about a point A is a vector and it is given by $\overrightarrow{AP} \times \overrightarrow{F}$, where P is a point on the line of the force $\overrightarrow{F}$.

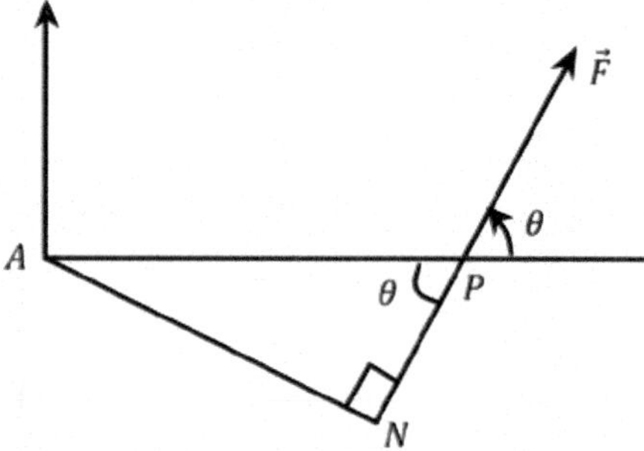

Figure 2.5 Represents the moment of a force about a point

Thus, the moment of the force $\overrightarrow{F}$ about a point A (See Figure 2.5) is a vector which is perpendicular to the plane containing $\overrightarrow{F}$ and A having a modulus

$$= \left|\overrightarrow{AP} \times \overrightarrow{F}\right| = AP\left|\overrightarrow{F}\right| \sin\theta = AN\left|\overrightarrow{F}\right| \ (\because AN = AP\sin\theta\,),$$

where AN is the line perpendicular to the line of the force $\overrightarrow{F}$.

Now let $\overrightarrow{F} = (F_1, F_2, F_3)\,,\ A(a_1, a_2, a_3)$ and $P(p_1, p_2, p_3)$. Then

$$\overrightarrow{AP} = \overrightarrow{OP} - \overrightarrow{OA} = (p_1 - a_1, p_2 - a_2, p_3 - a_3).$$

Moment of $\overrightarrow{F}$ about the point A

$$= \overrightarrow{AP} \times \overrightarrow{F}$$

$$= \begin{vmatrix} \hat{i} & \hat{j} & \hat{k} \\ p_1 - a_1 & p_2 - a_2 & p_3 - a_3 \\ F_1 & F_2 & F_3 \end{vmatrix}$$

$$= \{F_3\,(p_2 - a_2) - F_2\,(p_3 - a_3)\}\,\hat{i} - \{F_3\,(p_1 - a_1) - F_1\,(p_3 - a_3)\}\,\hat{j}$$
$$+ \{F_2\,(p_1 - a_1) - F_1\,(p_2 - a_2)\}\,\hat{k}$$

Note:

(1) The moment of the force $\overrightarrow{F}$ about a point on its line of action is always zero, because $\overrightarrow{AP}$ is a null vector.

(2) If concurrent forces $\overrightarrow{F_1}, \overrightarrow{F_2}, \overrightarrow{F_3}, \ldots$ act at the point P, then their moment about a point $A = \overrightarrow{AP} \times \overrightarrow{F}$, where $\overrightarrow{F} = \overrightarrow{F_1} + \overrightarrow{F_2} + \overrightarrow{F_3} + \ldots$

Illustration 2.19: A particle moves from the point $A\,(3, 2, -1)$ to the point $B\,(2, -1, 4)$ under the effect of force $\overrightarrow{F} = 4\hat{\imath} - 3\hat{\jmath} + 2\hat{k}$. Find the work done by the force.

Solution: Here, the force vector is $\overrightarrow{F} = 4\hat{\imath} - 3\hat{\jmath} + 2\hat{k} = (4, -3, 2)$ and the displacement is given by $\overrightarrow{AB} = (2, -1, 4) - (3, 2, -1) = (-1, -3, 5)$. So, the work done is obtained as follow:

$$\begin{aligned} \therefore \text{Work done by the force } \overrightarrow{F} &= \overrightarrow{F} \cdot \overrightarrow{AB} = (4, -3, 2) \cdot (-1, -3, 5) \\ &= -4 + 9 + 10 = 15 \text{ units} \end{aligned}$$

Note: In this illustration, the measurement unit for the force and displacement are not given so we have used a unit.

Illustration 2.20: The constant forces $\hat{\imath} + 2\hat{\jmath} + 3\hat{k}$ and $3\hat{\imath} + \hat{\jmath} + \hat{k}$ act on a particle. Under the action of these forces, a particle moves to $5\hat{\imath} + \hat{\jmath} + 2\hat{k}$ from $\hat{\jmath} - 2k$. Obtain total work done by these forces.

Solution: Here, $\overrightarrow{F_1} = \hat{\imath} + 2\hat{\jmath} + 3\hat{k} = (1, 2, 3)$ and $\overrightarrow{F_2} = 3\hat{\imath} + \hat{\jmath} + \hat{k} = (3, 1, 1)$ are the forces acting on the particle. So, the total force or resultant force acting on the particle is given by

$$\therefore \text{Resultant force } \overrightarrow{F} = \overrightarrow{F_1} + \overrightarrow{F_2} = (1, 2, 3) + (3, 1, 1) = (4, 3, 4)$$

Let A represent the initial position of the particle and B represents the final position of the particle after forces are applied. Therefore, $A = \hat{\jmath} - 2\hat{k} = (0, 1, -2)$ and $B = 5\hat{\imath} + \hat{\jmath} + 2\hat{k} = (5, 1, 2)$. Now, the displacement is given by

$$\therefore \text{Displacement } \overrightarrow{AB} = (5, 1, 2) - (0, 1, -2) = (5, 0, 4)$$

As we have obtained the total forces acting on the particle and the displacement of the particle, we calculate work done as follow:

$$\begin{aligned} \therefore \text{Work done} = \overrightarrow{F} \cdot \overrightarrow{AB} &= (4, 3, 4) \cdot (5, 0, 4) \\ &= 4 \times 5 + 3 \times 0 + 4 \times 4 \\ &= 20 + 0 + 16 = 36 \end{aligned}$$

Thus, the total work done is by the forces is 36 units.

Illustration 2.21: Forces $3\hat{\imath} - \hat{\jmath} + 2\hat{k}$ and $\hat{\imath} + 3\hat{\jmath} - \hat{k}$ act on a particle and the particle moves from $2\hat{\imath} + 3\hat{\jmath} + \hat{k}$ to $5\hat{\imath} + 2\hat{\jmath} + \hat{k}$ under these forces. Find the work done by these forces.

Solution: The given forces acting on a particle are $\overrightarrow{F_1} = 3\hat{\imath} - \hat{\jmath} + 2\hat{k} = (3, -1, 2)$ and $\overrightarrow{F_2} = \hat{\imath} + 3\hat{\jmath} - \hat{k} = (1, 3, -1)$.

The total force acting on a particle is given by

$$\therefore \text{Resultant force } \overrightarrow{F} = \overrightarrow{F_1} + \overrightarrow{F_2} = (3, -1, 2) + (1, 3, -1) = (4, 2, 1)$$

Let A represent the initial position of the particle then $A = 2\hat{\imath} + 3\hat{\jmath} + \hat{k}$ = (2,3,1) and let B denote the final position of the particle after forces are applied then $B = 5\hat{\imath} + 2\hat{\jmath} + \hat{k} = (5, 2, 1)$. So, the displacement of the particle from the point A to point B is given by

Displacement $= \overrightarrow{AB} = (5, 2, 1) - (2, 3, 1) = (3, -1, 0)$

The work done by the forces is obtained as follow:

$$\begin{aligned} \text{Work done} &= \overrightarrow{F} \cdot \overrightarrow{AB} \\ &= (4, 2, 1) \cdot (3, -1, 0) \\ &= 4 \times 3 + 2 \times (-1) + 1 \times 0 \\ &= 12 - 2 + 0 = 10 \text{ units.} \end{aligned}$$

Illustration 2.22: A particle moves from point $3\hat{\imath} - 2\hat{\jmath} + \hat{k}$ to point $\hat{\imath} + 3\hat{\jmath} - 4\hat{k}$ under the effect of constant forces $\hat{\imath} - \hat{\jmath} + \hat{k}$, $\hat{\imath} + \hat{\jmath} - 3\hat{k}$, $4\hat{\imath} + 5\hat{\jmath} - 6\hat{k}$. Find the work done by these forces on the particle.

Solution: Let A be the initial point of the position of the particle then $A = 3\hat{\imath} - 2\hat{\jmath} + \hat{k} = (3, -2, 1)$ and let B be the final position of the particle then $B = \hat{\imath} + 3\hat{\jmath} - 4\hat{k} = (1, 3, -4)$. The displacement of the particle is obtained as

Displacement $= \overrightarrow{AB} = (1, 3, -4) - (3, -2, 1) = (-2, 5, -5)$.

The constant forces acting on the particle are $\overrightarrow{F_1} = \hat{\imath} - \hat{\jmath} + \hat{k} = (1, -1, 1)$, $\overrightarrow{F_2} = \hat{\imath} + \hat{\jmath} - 3\hat{k} = (1, 1, -3)$, and $\overrightarrow{F_3} = 4\hat{\imath} + 5\hat{\jmath} - 6\hat{k} = (4, 5, -6)$.

The resultant forces acting on the particle is given by

$$\begin{aligned} \therefore \text{Resultant forces } \overrightarrow{F} &= \overrightarrow{F_1} + \overrightarrow{F_2} + \overrightarrow{F_3} \\ &= (1, -1, 1) + (1, 1, -3) + (4, 5, -6) \\ &= (6, 5, -8) \\ \therefore \text{Work done} &= \overrightarrow{F} \cdot \overrightarrow{AB} = (6, 5, -8) \cdot (-2, 5, -5) \\ &= -12 + 25 + 40 = 53 \text{ units} \end{aligned}$$

Illustration 2.23: Three forces of magnitudes $2, 1$, and 5 units in the directions $(1, 2, 3)$, $(-1, 2, 3)$, and $(-1, 2, -3)$ are applied on a particle. If a particle moves from the point $(0, 1, -2)$ to the point $(-1, 3, 2)$ under these forces, find the work done.

Solution: Let $\vec{x} = (1, 2, 3)$, $\vec{y} = (-1, 2, 3)$, and $\vec{z} = (-1, 2, -3)$ are the given directions, and to find the forces in these directions, we have to find the unit vectors in these directions.

$$\therefore |\vec{x}| = \sqrt{1+4+9} = \sqrt{14}, \ |\vec{y}| = \sqrt{1+4+9} = \sqrt{14},$$

and similarly $|\vec{z}| = \sqrt{14}$.

$$\therefore \text{The unit vector in the direction of } \vec{x} = \frac{\vec{x}}{|\vec{x}|} = \frac{1}{\sqrt{14}}(1, 2, 3)$$

$$\therefore \text{The unit vector in the direction of } \vec{y} = \frac{\vec{y}}{|\vec{y}|} = \frac{1}{\sqrt{14}}(-1, 2, 3)$$

$$\therefore \text{The unit vector in the direction of } \vec{z} = \frac{\vec{z}}{|\vec{z}|} = \frac{1}{\sqrt{14}}(-1, 2, -3)$$

$$\therefore \quad \vec{F_1} = \text{Force of magnitude 2 in the direction of}$$

$$\vec{x} = \frac{2}{\sqrt{14}}(1, 2, 3)$$

$$\therefore \quad \vec{F_2} = \text{Force of magnitude 1 in the direction of}$$

$$\vec{y} = \frac{1}{\sqrt{14}}(-1, 2, 3)$$

$$\therefore \quad \vec{F_3} = \text{Force of magnitude 5 in the direction of}$$

$$\vec{z} = \frac{5}{\sqrt{14}}(-1, 2, -3)$$

$$\therefore \text{The resultant force } \vec{F} = \vec{F_1} + \vec{F_2} + \vec{F_3}$$

$$= \frac{1}{\sqrt{14}}[2\,(1, 2, 3) + (-1, 2, 3) + 5(-1, 2, -3)]$$

$$= \frac{1}{\sqrt{14}}\,[(2, 4, 6) + (-1, 2, 3) + (-5, 10, -15)]$$

$$= \frac{1}{\sqrt{14}}(-4, 16, -6)$$

Now, let $A(0, 1, -2)$ be the initial position of the particle and $B\,(-1, 3, 2)$ be the final position of the particle then the displacement of the particle is given by

Displacement $= \overrightarrow{AB} = (-1, 3, 2) - (0, 1, -2) = (-1, 2, 4)$.

Finally, the work done is given as

$$\therefore \text{Work done} = \vec{F} \cdot \overrightarrow{AB} = \frac{1}{\sqrt{14}}(-4, 16, -6) \cdot (-1, 2, 4)$$
$$= \frac{1}{\sqrt{14}}(4 + 32 - 24) = \frac{12}{\sqrt{14}}$$

Thus, the total work done is obtained as $\frac{12}{\sqrt{14}}$ units.

Illustration 2.24: Find the moment about the point $(2, 3, -1)$ of the force $3\hat{i} - \hat{k}$ acting through the point $(1, -2, 1)$. Also, find the magnitude of the moment.

Solution: Let $A(2, 3, -1)$ and $P(1, -2, 1)$ be the given points.

Let $\vec{F} = 3\hat{i} - \hat{k} = (3, 0, -1)$ be the given force vector.

$$\therefore \overrightarrow{AP} = \overrightarrow{OP} - \overrightarrow{OA} = (1, -2, 1) - (2, 3, -1) = (-1, -5, 2)$$

Moment of the force $\vec{F}$ about the point $A = \overrightarrow{AP} \times \vec{F}$

$$= \begin{vmatrix} \hat{i} & \hat{j} & \hat{k} \\ -1 & -5 & 2 \\ 3 & 0 & -1 \end{vmatrix}$$
$$= (5 - 0)\,\hat{i} - (1 - 6)\,\hat{j} + (0 + 15)\,\hat{k}$$
$$= 5\hat{i} + 5\hat{j} + 15\hat{k} = (5, 5, 15)$$

$$\therefore \text{Magnitude of the moment} = \sqrt{25 + 25 + 225} = \sqrt{275} = 5\sqrt{11}.$$

Illustration 2.25: Find the moment about the point $(4, 0, 1)$ of the forces $2\hat{i} + \hat{j} - 3\hat{k}$ and $2\hat{i} - 2\hat{j} + \hat{k}$ acting through the point $(-1, 3, -2)$.

Solution: The resultant force $\vec{F}$ is given by

$$\vec{F} = 2\hat{i} + \hat{j} - 3\hat{k} + 2\hat{i} - 2\hat{j} + \hat{k} = (4, -1, -2).$$

Let $A(4, 0, 1)$ and $P(-1, 3, -2)$ be the given points.

$$\therefore \overrightarrow{AP} = (-1, 3, -2) - (4, 0, 1) = (-5, 3, -3)$$

The moment of the force $\overrightarrow{F}$ about the point A

$$= \overrightarrow{AP} \times \overrightarrow{F} = \begin{vmatrix} \hat{i} & \hat{j} & \hat{k} \\ -5 & 3 & -3 \\ 4 & -1 & -2 \end{vmatrix}$$

$$= (-6 - 3)\,\hat{i} - (10 + 12)\,\hat{j} + (5 - 12)\,\hat{k}$$

$$= -9\hat{i} - 22\hat{j} - 7\hat{k}$$

2.6 Exercise

1. Evaluate $\left(\hat{i} + 2\hat{j} + \hat{k}\right) \cdot \left(3\hat{k} - 2\hat{j} + 4\hat{i}\right)$.

   (Answer : 3)

2. If $\overrightarrow{a} = \hat{i} - \hat{j} + \hat{k}$, $\overrightarrow{b} = 2\hat{i} - \hat{j} + \hat{k}$, and $\overrightarrow{c} = \hat{i} + \hat{j} - 2\hat{k}$, then find $\overrightarrow{a} \cdot \left(\overrightarrow{b} + \overrightarrow{c}\right)$.

   (Answer : 2)

3. For what value of p, the vectors $2\hat{i} + 3\hat{j} - \hat{k}$ and $p\hat{i} - \hat{j} + 3\hat{k}$ are perpendicular to each other?

   (Answer :p= 3)

4. If the vectors $\overrightarrow{a} = m\hat{i} - 2\hat{j} + \hat{k}$ and $\overrightarrow{b} = 2m\hat{i} + m\hat{j} - 4\hat{k}$ are perpendicular to each other then find the value of m.

   (Answer : $m = -1,\ 2$)

5. If vectors $(m, 2m, 4)$ and $(m,\ -3, 2)$ are mutually perpendicular then find the value of m.

   (Answer : $m = 2,\ 4$)

6. Find the angle between vectors $(1, 2, 3)$ and $(-2, 3, 1)$.

   (Answer :$\frac{\pi}{3}$ or 60°)

7. For which value of p vectors $\overrightarrow{x} = (1, -2, -3)$ and $\overrightarrow{y} = (2, p, 4)$ are mutually perpendicular?

   (Answer : $p = -5$)

8. Find the value of q if vectors $\overrightarrow{a} = (q, 2, 1)$, and $\overrightarrow{b} = (2, q, -4)$ are perpendicular to each other.

(Answer : $q = 1$)

9. Find the modulus of $\left(2\hat{\imath} - 3\hat{\jmath} + \hat{k}\right) \times (\hat{\imath} - \hat{\jmath} + 2\hat{k})$.

(Answer :$\sqrt{35}$)

10. Find a unit perpendicular vector to the given vector $\overrightarrow{x} = (1,\ 2,\ 3)$ and $\overrightarrow{y} = (-2,\ 1, -2)$.

$\left(\text{Answer : } \frac{1}{3\sqrt{10}}(-7,\ -4,\ 5)\right)$

11. Find a unit perpendicular vector to the given vector $\overrightarrow{a} = (1, -1, 1)$ and $\overrightarrow{b} = (2, 3, -1)$ both.

$\left(\text{Answer : } \frac{1}{\sqrt{38}}(-2,\ 3,\ 5)\right)$

12. Find a unit vector perpendicular to the plane containing the vectors $\overrightarrow{a} = 3\hat{\imath} + \hat{\jmath} + 2\hat{k}$ and $\overrightarrow{b} = 2\hat{\imath} + \hat{\jmath} - \hat{k}$.

$\left(\text{Answer : } \frac{1}{\sqrt{83}}(-3,\ 7,\ 5)\right)$

13. Show that the angle between the vectors $\hat{\imath} + \hat{\jmath} - \hat{k}$ and $2\hat{\imath} - \widehat{2\jmath} + \hat{k}$ is $\sin^{-1}\sqrt{\frac{26}{27}}$.

14. Prove that the angle between two vectors $3\hat{\imath} + \hat{\jmath} + 2\hat{k}$ and $2\hat{\imath} - 2\hat{\jmath} + 4\hat{k}$ is $\sin^{-1}\frac{2}{\sqrt{7}}$.

15. Prove that $|\overrightarrow{x} + \overrightarrow{y}|^2 + |\overrightarrow{x} - \overrightarrow{y}|^2 = 2(|\overrightarrow{x}|^2 + |\overrightarrow{y}|^2)$.

16. If $\overrightarrow{x} + \overrightarrow{y} + \overrightarrow{z} = \theta$, then prove that $\overrightarrow{x} \times \overrightarrow{y} = \overrightarrow{y} \times \overrightarrow{z} = \overrightarrow{z} \times \overrightarrow{x}$.

17. If the vectors $\hat{\imath} - 2\hat{\jmath} - 3\hat{k}$ and $\hat{\imath} + 2\hat{k}$ represent two sides of the triangle then find the area of the triangle.

(Answer : $\frac{1}{2}\sqrt{45}$)

18. Find the area of the parallelogram whose adjacent sides are $3\hat{\imath} + \hat{\jmath} - 2\hat{k}$ and $\hat{\imath} - 3\hat{\jmath} + 4\hat{k}$.

(Answer : $10\sqrt{3}$)

19. If $|x - y| = 13$ for the vectors $\overrightarrow{x} = (7, p+1, 1)$ and $\overrightarrow{y} = (4, 1, 5)$, then find the value of p.

(Answer : -21)

20. Constant forces $(3, -2, 1)$ and $(-1, -1, 2)$ acts on the particle. Under the effects of these forces, a particle is displaced from $(2, 2, -3)$ to the point $(-1, 2, 4)$. Find the total work done by these forces.

(Answer : 15 unit)

21. Under the effect of two forces $(4, 1, -3)$ and $(3, 1, -1)$ particle is displaced from origin to $(1, 1, 1)$. Find the work done.

(Answer : 5 unit)

22. A particle is displaced from point $(0, 1, -2)$ to the point $(5, 1, 2)$ under the effect of constant forces $(1, 2, 3)$ and $(3, 1, 1)$ then find the total work done.

(Answer : 36 unit)

23. A particle moved from the point $\hat{i} - \hat{j}$ to point $3\hat{i} + \hat{k}$ under the effect of the two constant forces $-2\hat{k} + \hat{i} + \hat{j}$ and $2\hat{j} + 2\hat{i} - 4\hat{k}$. Then find the total work done.

(Answer : 3 unit)

24. Force $\hat{j} + \hat{i} + \hat{k}$ is action at $-2\hat{i} + 3\hat{j} + 4\hat{k}$ find the moment of inertia and its magnitude above $2\hat{i} + 3\hat{j} + 5\hat{k}$.

(Answer : $(1, 3, -4)$ and $\sqrt{26}$)

25. Force $(1, 1, 1)$ is acting on $B(1, 2, 3)$ find the magnitude of the moment of inertia along $A(-1, 2, 0)$.

(Answer : $(-3, 1, 2)$ and $\sqrt{14}$)

3

Vector Differential Calculus

3.1 Introduction

In this chapter, we study vector functions in three dimensions and the applications of differential calculus to them. Vectors simplify many calculations considerably and help to visualize physical and geometrical quantities and relations between them. Consequently, vector methods are used extensively in applied mathematics and engineering. The impact of these methods on the study of physical phenomena such as fluid flow, elasticity, heat flow, electrostatics, electromagnetism, and waves in solids and fluids, which the engineer must understand as the foundation for the design and construction of systems such as aircraft, laser generators, robots, and thermo-dynamical systems, is critical to the engineer.

Our goal is to acquaint readers with vector calculus, a branch of differential calculus that applies the basic notions of ordinary differential calculus to vector functions. The gradient, divergence, and curl are three physically and geometrically essential concepts connected to scalar and vector fields.

3.2 Vector and Scalar Functions and Fields

A variable quantity whose value at any point in a region of space depends on the position of the point is called a point function. There are two types of point functions. A point function whose values are vector is called vector functions and is given as

$$\vec{v} = \vec{v}\,(P) = \left(v_1\,(P)\,,\; v_2\,(P)\,, v_3(P)\right)$$

depending on the position of the points P in space. A point function whose values are scalars at any point P is called scalar functions and is given as $f = f(P)$ depending on the position of the point P in space.

The domain of definition for such a function in applications is a region of space, a surface in space, or a curve in space.

3.2.1 Scalar Function and Field

If to each point (x, y, z) of a region R in space corresponds a number or a scalar $f = f(x, y, z)$ then f is called a scalar function of position or scalar (point) function and R is called a scalar field.

For example, scalar functions are defined as quantities that take distinct values at various sites, such as the temperature field of a body, the pressure field of air in the earth's atmosphere, density of a body, and potential. $\phi(x, y, z) = x^2 - 2yz^3$ defines a scalar field. A scalar field that is independent of time is called a stationary or steady-state scalar field.

3.2.2 Vector Function and Field

If to each point (x, y, z) of a region R in space, there corresponds a vector $\vec{v} = \vec{v}(x, y, z) = [v_1(x, y, z), v_2(x, y, z), v_3(x, y, z)]$, then $\vec{v}$ is called a vector function of position or vector (point) function and R is called a vector field.

For example, vector functions include the velocity of a flowing fluid at any given time, gravitational force, electric field intensity, magnetic field intensity, and force. $\vec{v}(x, y, z) = xy^2\,\hat{i} - 2yz^2\hat{j} + x^2\hat{k}$ defines a vector field. A vector field that is independent of time is called a stationary or steady-state vector field.

3.2.3 Level Surfaces

Let a scalar function $f(x, y, z)$ be defined in a certain region R of space. Consider those points of the field for which scalar function f has a fixed value k. The totality of point satisfying the equation $f(x, y, z) = k$ defines in general a surface and is called a level surface of the function since at every point of the surface f has a constant value k. For different values of k, we have different level surfaces and there will be no intersection of two-level surfaces. For example, if $f(x, y, z)$ represents temperature in a medium, then $f(x, y, z) = k$ represents a surface on which the temperature is a constant k. These surfaces are called isothermal surfaces.

For example, the level surfaces of the scalar fields in space defined by the function $f(x, y, z) = x^2 + y^2 + z^2$ is $f(x, y, z) = k$ i.e., $x^2 + y^2 + z^2 = k$. Therefore, the level surfaces are spheres of radius $\sqrt{k}$.

3.3 Curve and Arc Length

First, we discuss the parametric representation of curves.

3.3.1 Parametric Representation of Curves

If x and y are given as continuous functions

$$x = f(t),\ y = g(t)$$

over an interval I of $t-$values, then the set of points $(x,\ y) = (f(t),\ g(t))$ defined by these equations is a curve in the coordinate plane. The equations are parametric equations for the curve. The variable t is a parameter for the curve and its domain I is the parameter interval. If I is a closed interval, $a \leq t \leq b$, the point $(f(a),\ g(a))$ is the initial point of the curve and $(f(b),\ g(b))$ is the terminal point of the curve. If the equations of a curve are given in the plane in parametric equations and parametric interval then it is called a parameterized curve. The equations and interval constitute a parameterization of the curve.

A curve C in the two-dimensional xy-plane can be parametrized by

$$x = x(t),\ y = y(t),\ a \leq t \leq b.$$

Then, the position vector of a point P on the C can be written as

$$\vec{r}(t) = x(t)\hat{\imath} + y(t)\hat{\jmath} \tag{3.1}$$

Therefore, the position vector of a point on a curve defines a vector function. Similarly, a three-dimensional curve or a space curve C can be parametrized as

$$\vec{r}(t) = x(t)\hat{\imath} + y(t)\hat{\jmath} + z(t)\hat{k},\ a = t = b \tag{3.2}$$

Illustration 3.1: Find a parametrization for the line segment with endpoints $(-2,\ 1)$ and $(3,\ 5)$.

Solution:

Using $(-2,\ 1)$, we crate the parametric equations $x = -2 + at,\ y = 1 + bt$.

These represent a line, as we see by solving each equation for t and equating to obtain

$$\frac{x+2}{a} = \frac{y-1}{b}$$

This line goes through the point $(-2,\ 1)$ when $t = 0$. We determine a and b so that the line goes through $(3,\ 5)$ when $t = 1$.

$$3 = -2 + a \Rightarrow a = 5x = 3\ when\ t = 1$$
$$5 = 1 + b \Rightarrow b = 4y = 5\ when\ t = 1$$

Therefore,

$$x = -2 + 5t,\ \ y = 1 + 4t,\ 0 = t = 1.$$

or

$$\vec{r}\ (t) = (-2 + 5t)\ \hat{i} + (1 + 4t)\ \hat{j}$$

is a parametrization of the line segment with initial point $(-2,\ 1)$ and terminal point $(3,\ 5)$.

Illustration 3.2: Find the parametric representation of the straight line through the point $P\,(1,\ 2,\ 3)$ and has the direction $b = i + 2j + 2k$.

Solution:

In this case, $a = (1 - 0)\ i + (2 - 0)\ j + (3 - 0)\ k = i + 2j + 3k$

$$\therefore a_1 = 1,\ a_2 = 2,\ a_3 = 3$$

from $b = i + 2j + 2k$, we have $b_1 = 1,\ b_2 = 2,\ b_3 = 2$

$$\therefore \vec{r}\ (t) = a + tb = (a_1 + tb_1)\,\hat{i} + (a_2 + tb_2)\,\hat{j} + (a_3 + tb_3)\,\hat{k}$$

gives $\vec{r}\ (t) = (1 + t)\,\hat{i} + (2 + 2t)\,\hat{j} + (3 + 2t)\,\hat{k}$

or $\vec{r}\ (t) = (1 + t)\,\hat{i} + 2\,(1 + t)\,\hat{j} + (3 + 2t)\,\hat{k}$ are required for parametric representation.

3.3.2 Curves with Tangent Vector

A curve is the locus of a point whose position vector $\vec{r}$ relative to a fixed origin may be expressed as a function of a single variable parameter. Then its Cartesian coordinates $x,\ y,\ z$ are also functions of the same parameter.

Thus, the equation of the curve is usually expressed in parametric form

$$x = x\,(t)\,,\ y = y\,(t)\,,\ z = z\,(t) \tag{3.3}$$

If $P,\ Q$ are adjoining points $(x,\ y,\ z)$ and $(x + \delta x,\ y + \delta y,\ z + \delta z)$, then their position vectors are given by (Figure 3.1)

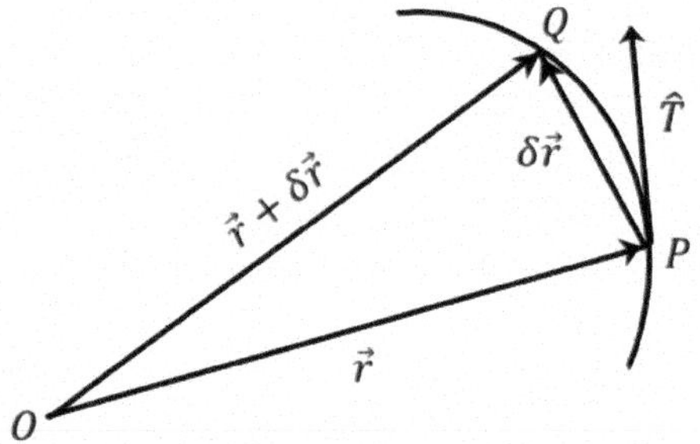

Figure 3.1 Represents a curve with a tangent vector

$$\vec{r} = x\hat{i} + y\hat{j} + z\hat{k} \text{ and } \vec{r} + \delta\vec{r} = (r\,\delta x)\,\hat{i} + (y + \delta y)\,\hat{j} + (z + \delta z)\,\hat{k}$$

$$\therefore \delta\vec{r} = (\delta x)\,\hat{i} + (\delta y)\,\hat{j} + (\delta z)\,\hat{k}$$

$$\therefore |d\vec{r}|^2 = \delta r^2 = \delta x^2 + \delta y^2 + \delta z^2 \tag{3.4}$$

If δs is the length of the are PQ and in the limit as $\delta t \to 0,$ chord $PQ\ (= \delta r)$ and are δs will be equal i.e., $\frac{dr}{ds} = 1$

$$\therefore \left(\frac{d\vec{r}}{ds}\right)^2 = \left(\frac{dx}{ds}\right)^2 + \left(\frac{dy}{ds}\right)^2 + \left(\frac{dz}{ds}\right)^2 = 1 \tag{3.5}$$

and

$$\frac{d\vec{r}}{ds} = \frac{dx}{ds}\,\hat{i} + \frac{dy}{ds}\,\hat{j} + \frac{dz}{ds}\hat{k}$$

$$\therefore \left|\frac{d\vec{r}}{ds}\right| = \sqrt{\left(\frac{dx}{ds}\right)^2 + \left(\frac{dy}{ds}\right)^2 + \left(\frac{dx}{ds}\right)^2} = 1 \quad (\because \text{ from } (3.4)) \tag{3.6}$$

3.3.2.1 Tangent Vector

Limiting position of chord PQ (i.e., $d\vec{r}$) as $\delta t \to 0$ (as $Q \to P$) is along the tangent to the curve at P

$$\frac{d\vec{r}}{ds} = \left|\frac{d\vec{r}}{ds}\right|\hat{T} = \hat{T} \qquad (\text{ from } (3.5))$$

where unit vector $\hat{T}$ is along the tangent at $P\,(t)$.

$$\therefore \frac{dr}{dt} = \frac{dr}{ds}\cdot\frac{ds}{dt} = \frac{ds}{dt}\cdot\hat{T} \tag{3.7}$$

where, $\hat{T}$ is a tangent vector.

3.3.2.2 Important Concepts

In Figure 3.2, the limiting position of the PQR passing through neighboring points P, Q, R on the curve and as Q, R approaches to P (i.e., the plane containing two consecutive tangents and therefore three consecutive points at P) is called the osculating plane or the plane of curvature of the curve at point P.

The binormal at the point P is denoted by $\hat{B}$ and is a line normal to the osculating plane at $P\left(i.e.,\ \hat{B}\perp\hat{T}\right)$. And the principal normal at the point P denoted by $\hat{n}$ to the curve at point P is a line through the point P lying in the osculating plane at P and it is perpendicular to the tangent line. $\hat{N}\perp\hat{T}$ and $\hat{N}\perp\hat{B}$.

The binormal is perpendicular to both $\hat{T}$ and $\hat{N}$, it is parallel to $\hat{T}\times\hat{N}$ Hence trio unit vectors $\hat{T},\ \hat{N},\ \hat{B}$ form a right-handed system of mutually perpendicular unit vectors, and therefore connected by the relations

$$\hat{T}\,.\,\hat{T}=\hat{N}\,.\,\hat{B}=\hat{B}\,.\,\hat{T}=0$$

and

$$\hat{T}\times\hat{N}=\hat{B}\,,\ \hat{N}\times\hat{B}=\hat{T}\,,\ \hat{B}\times\hat{T}=\hat{N},$$

the cyclic order being preserved in the cross products.

(a) A plane through the point P, normal to $\hat{B}$ is an osculating plane of the curve at P.

(b) A plane through the point P normal to $\hat{T}$ is known as the normal plane of the curve at P.

(c) A plane through the point P normal to $\hat{N}$ is known as rectifying plane of the curve.

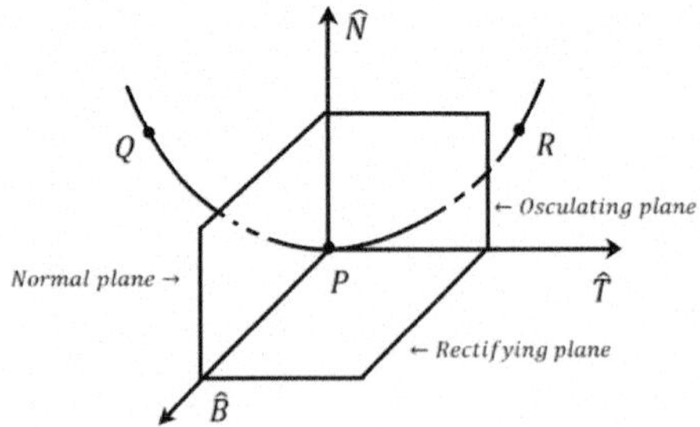

Figure 3.2 Represents the plane of curvature of the curve

3.3.3 Arc Length

The parameterization for a curve is a set of functions depending only on a parameter t along with the bounds for the parameter. When we parameterize a curve by taking values of t from some interval $[a,\ b]$, the position vector $\vec{r}(t)$ of any point, t on the curve can be written as,

$$\vec{r}\,(t) = x\,(t)\,\hat{i} + y\,(t)\,\hat{j} + z\,(t)\,\hat{k}$$

The tangent vector $\vec{r}\,'(t)$ is

$$\vec{r}\,'\,(t) = x'\,(t)\,\hat{i} + y'\,(t)\,\hat{j} + z'\,(t)\,\hat{k}$$

$$\therefore \left|\vec{r}\,'\,(t)\right| = \sqrt{(x'\,(t))^2 + (y'\,(t))^2 + (z'\,(t))^2}$$

The length of the curve is

$$l = \int_a^b \left|\vec{r}\,'\,(t)\right| dt$$

The arc length function or arc length of the curve is obtained by considering a variable limit. i.e., $a = t_1$ and $b = t_2$.

$$l = \int_{t_1}^{t_2} \left|\vec{r}\,'\,(v)\right| dv \tag{3.8}$$

3.3.3.1 Unit Tangent Vector

The unit tangent vector of a curve $\vec{r}\,(t)$ is

$$\hat{T} = \frac{d\vec{r}}{ds} = \frac{d\vec{r}/dt}{ds/dt} = \frac{v}{|v|} \tag{3.9}$$

Illustration 3.3: Find the length of the arc between points $(a,\ 0,\ 0)$ and $(0,\ a,\ \frac{1}{2}\pi a\tan\alpha\ )$ for the curve $x = a\cos\theta\ ,\ y = a\sin\theta\ ,\ z = a\,\theta\tan\alpha\ .$

Solution:

Using result (3.8), we get

$$\left(\frac{ds}{d\theta}\right)^2 = \left(\frac{dx}{d\theta}\right)^2 + \left(\frac{dy}{d\theta}\right)^2 + \left(\frac{dz}{d\theta}\right)^2$$
$$= a^2\sin^2\theta + a^2\cos^2\theta + a^2\tan^2\alpha$$
$$= a^2(1 + \tan^2\alpha) = a^2\sec^2\alpha$$
$$\therefore\ \frac{ds}{d\theta} = a\ \sec\alpha$$

For the given points the parameters are
$\theta = 0[a,\ 0,\ 0]$ and $\theta = \frac{\pi}{2}[0,\ a,\ \frac{1}{2}\ \pi a\ \tan\alpha]$
$\therefore$ The required length $= \int_0^{\pi/2} a\ \sec\alpha \cdot d\theta = \frac{\pi}{2}\ a\ \sec\alpha$.

Illustration 3.4: Prove that the length of the curve
$x = 2a\left(\sin^{-1}t\ + t\sqrt{1-t^2}\ \right),\ y = 2\ at^2,\ z = 4$ at
between the points, $t = t_1$ and $t = t_2$ are $4\sqrt{2}\ (t_2 -\ t_1)\,a$.

Solution:
Now,

$$\frac{dx}{dt} = 2a\left[\frac{1}{\sqrt{1-t^2}} + \sqrt{1-t^2} - 1\left(-\frac{2t}{2\sqrt{1-t^2}}\right)\right]$$

$$= 2a\left[\frac{1-t^2}{\sqrt{1-t^2}} + \sqrt{1-t^2}\right] = 4a\sqrt{1-t^2}$$

$$\frac{dy}{dt} = 4\ at,\ \frac{dz}{dt} = 4a$$

$$\therefore \left(\frac{ds}{dt}\right)^2 = \left(\frac{dx}{dt}\right)^2 + \left(\frac{dy}{dt}\right)^2 + \left(\frac{dz}{dt}\right)^2$$

$$= 16a^2\left(1-t^2\right) + 16a^2t^2 + 16a^2$$

$$= 16^2\left[1-t^2\right) + t^2 + 1] = 32a^2$$

$$\therefore\ \frac{ds}{dt} = 4\sqrt{2a}$$

$$\therefore s = 4\sqrt{2a}\int_{t_1}^{t_2} dt = 4\sqrt{2}\,a\ (t_2 - t_1)$$

Illustration 3.5: Find the length of the arc of the curve
$y =$ log sec x from $x = 0$ to $x = \pi/3$

Solution:
Here,

$$\frac{dy}{dx} = \frac{1}{\sec x} \cdot \sec x\ \tan x\ = \tan x$$

Length of the arc $S = \int_0^{\pi/3} \sqrt{\left[1 + \left(\frac{dy}{dx}\right)^2\right]}\ dx$

$$= \int_0^{\pi/3} \sqrt{1 + \tan^2 x} \; dx$$

$$= \int_0^{\pi/3} \sec x \, dx = [\log(\sec x - \tan x)]_0^{\pi/3}$$

$$= \log \left[\sec\left(\frac{\pi}{3}\right) - \tan\left(\frac{\pi}{3}\right) \right] - \log(1 - 0)$$

$$= \log(2 - \sqrt{3}) - 0$$

$$= \log(2 - \sqrt{3})$$

Illustration 3.6: Find the arc length of the Helix traced by

$$\vec{r}\,(t) = a \cos t \; \hat{i} + a \sin \hat{j} + ct \, \hat{k}, \; a > 0, \;\; 0 \leq t \leq 2\pi \, .$$

Also, express the position vector $\vec{r}\,(t)$ in terms of the arc length s.

Solution:

we have $x\,(t) = a \cos t$, $y\,(t) = a \sin t$, $z\,(t) = ct$ and

$$\frac{dx}{dt} = -a \sin t \, , \quad \frac{dy}{dt} = a \cos t \, , \quad \frac{dz}{dt} = c$$

Therefore,

$$s = \text{Arc length} = \int_0^{2\pi} [a^2 \sin^2 t \; + a^2 \cos^2 t \; + c^2]^{1/2}$$

$$= \int_0^{2\pi} \left(a^2 + c^2\right)^{1/2} \; dt \; = 2\pi \left(a^2 + c^2\right)^{1/2}$$

Also, the arc length s is given as

$$s = \int_0^t \left(a^2 + c^2\right)^{1/2} dt = t\left(a^2 + c^2\right)^{1/2}$$

or

$$t = \frac{s}{\left(a^2 + c^2\right)^{1/2}}$$

Therefore,

$$r\,(s) = a \cos\left(\frac{s}{\sqrt{(a^2 + c^2)}}\right) i + a \sin\left(\frac{s}{\sqrt{(a^2 + c^2)}}\right) j$$

$$+ c\left(\frac{s}{\sqrt{(a^2 + c^2)}}\right) k \, .$$

It can be verified that $\left|\frac{dr}{ds}\right| = 1.$

Illustration 3.7: Find the unit tangent vector to the unit circle

$$\vec{r}\,(t) = (\cos t)\,\hat{\imath} + (\sin t)\,\hat{\jmath}$$

Solution:

Here, $r\,(t) = (\cos t)\,\hat{\imath} + (\sin t)\,\hat{\jmath}$

$$\therefore \vec{v} = \frac{dr}{dt} = (-\sin t)\,\hat{\imath} + (\cos t)\,\hat{\jmath}$$

$$\left|\sqrt{\vec{v}}\right| = 1$$

$$\therefore \hat{T} = \frac{\vec{v}}{|\vec{v}|} = \frac{(-\sin t\,)\,\hat{\imath} + (\cos t)\,\hat{\jmath}}{1} = (-\sin t)\,\hat{\imath} + (\cos t)\,\hat{\jmath}$$

3.4 Curvature and Torsion

In this section, we discuss curvature, torsion, and Serret (1851) – Frenet (1852) Formulas. We may gather them together in the form.

(a) $\frac{d\hat{T}}{dS} = k\hat{N}$, (b) $\frac{d\hat{B}}{dS} = -\tau\hat{N}$, (c) $\frac{d\hat{N}}{dS} = \tau\hat{B} - k\hat{N}$

These formulas give the derivatives of the direction cosines of the tangent, the binomial, and the principal normal.

(I) Definition: Curvature k at point P on the curve is are the rate of rotation of tangent $\hat{T}$ at P

$$\therefore k = \frac{d\Psi}{dS} \tag{i}$$

where $\delta\Psi$ is the angle through which $\hat{T}$ has turned when P moves to a point Q. $\frac{d\hat{T}}{dS} = \frac{d\hat{T}}{d\Psi} \cdot \frac{d\Psi}{ds} = k\frac{d\hat{T}}{d\Psi}$ (ii)

Now,

$$\hat{T}\,.\hat{T} = 1$$

$$\therefore\ \hat{T} \cdot \frac{d\hat{T}}{dS} = 0 \Rightarrow \hat{T} \perp \frac{d\hat{T}}{dS}$$

Since $\hat{T}$ and $\hat{T} + \delta\hat{T}$ lie in the osculating plane, $\delta\hat{T}$ lies in the osculating plane i.e., $\frac{d\hat{T}}{dS}$ lies in an osculating plane and being perpendicular to, it is

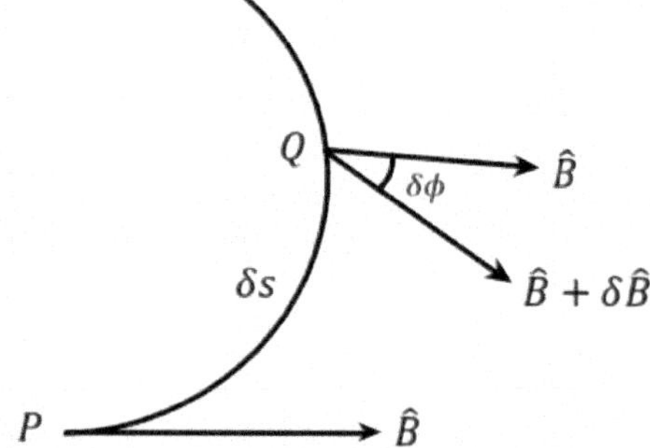

Figure 3.3 Represents the arc rate of rotation of binormal

along $\hat{N}$ at P.

$$\therefore \frac{d\hat{T}}{d\psi} = (1)\ \hat{N} \qquad \left(\because \left|\frac{d\hat{T}}{D\psi}\right| = 1\right) \tag{iii}$$

From (ii) and (iii) we get,

$$\frac{d\hat{T}}{dS} = k\hat{N} \tag{3.10}$$

(II) Definition: Torsion (τ) at point P on the curve is arc-rate of rotation of binormal at P.

If $\delta\emptyset$ is the angle of rotation when P moves to point Q (in Figure 3.3).

$$\therefore \tau = \frac{d\phi}{dS} \tag{i}$$

$$\frac{d\hat{B}}{dS} = \frac{d\hat{B}}{d\phi} \cdot \frac{d\phi}{ds} = \tau\ \frac{d\hat{B}}{d\emptyset} \tag{ii}$$

Now, $\hat{B} \cdot \hat{B} = 1, \ \therefore\ \hat{B} \cdot \frac{d\hat{B}}{d\phi} = 0$

$$\therefore \hat{B} \perp \frac{dB}{d\phi}$$

Since $\hat{B}$ and $\hat{B} + \delta\hat{B}$ (in limiting position as $\rightarrow 0$) lie in a normal plane, $\frac{d\hat{B}}{d\phi}$ lies in a normal plane and is perpendicular to $\hat{B}$. Hence $\frac{d\hat{B}}{d\phi}$ is along the normal $\hat{N}$ at P.

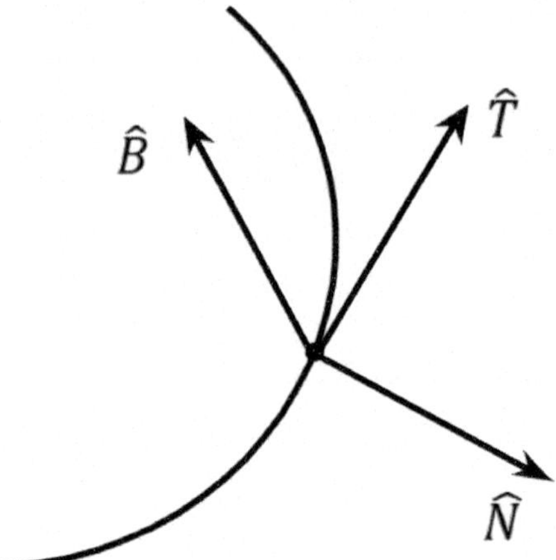

Figure 3.4 Represents $\hat{B}$, $\hat{T}$, and $\hat{N}$ orthogonal unit vectors

If point P moves in the increasing direction of s and the right-handed screw is rotated in such a way as to advance in direction of s (i.e., $\hat{T}$), the vector $\hat{B}$ is rotated in the clockwise direction and thus the direction of $\frac{d\hat{B}}{ds}$ is opposite to that of $\hat{N}$.

$$\therefore \frac{d\hat{B}}{d\emptyset} = -\hat{N} \left(\because \left| \frac{d\hat{B}}{d\phi} \right| = 1 \right) \quad \text{(iii)}$$

From (ii) and (iii), we get

$$\frac{d\hat{B}}{ds} = \tau \frac{d\hat{B}}{d\phi} = -\tau \hat{N}$$

Hence,

$$\frac{d\hat{B}}{ds} = -\tau \hat{N} \quad (3.11)$$

(III) Since $\hat{B}$, $\hat{T}$, and $\hat{N}$ are orthogonal unit vectors (in Figure 3.4),

$$\hat{N} = \hat{B} \times \hat{T}$$

$$\therefore \frac{d\hat{N}}{ds} = \hat{B} \times \frac{d\hat{T}}{ds} + \frac{d\hat{B}}{ds} \times \hat{T}$$

$$= \hat{B} \times \left(k\,\hat{N}\right) + \left(-\tau \hat{N}\right) \times \hat{T}$$

Using results (3.10) and (3.11), we get

$$= k\left(\hat{B}\times\hat{N}\right)-\tau\left(\hat{N}\times\hat{T}\right)$$

$$= k\left(--\hat{T}\right)-\tau\left(-\hat{B}\right)$$

Hence,

$$\frac{d\hat{N}}{ds} = \tau\hat{B}-k\hat{T} \tag{3.12}$$

3.4.1 Formulas for Curvature and Torsion

The position vector $\vec{r}\,(t)$ of the point $[x\,(t)\ ,\ y\,(t)\,,\ z\,(t)]$ on the curve is given by

$$\vec{r}\,(t) = x\,(t)\ \hat{\imath}+y\,(t)\ \hat{\jmath}+z\,(t)\ \hat{k} \tag{i}$$

where $x = x\,(t)\,,\ y = y\,(t)\,,\ z = z\,(t)$ represents parametric equations of the curve $\vec{r} = \vec{r}\,(t)$

$$\therefore\ \dot{\vec{r}} = \frac{d\vec{r}}{dt} = \frac{d\vec{r}}{ds}\cdot\frac{ds}{dt} = \frac{ds}{dt}\cdot\hat{T} \tag{ii}$$

$$\ddot{\vec{r}} = \frac{d^2r}{dt^2} = \frac{d^2s}{dt^2}\,\hat{T}+\frac{ds}{dt}\cdot\frac{d\hat{T}}{dt}$$

$$= \frac{d^2s}{dt^2}\ \hat{T}+\left(\frac{ds}{dt}\right)\frac{d\hat{T}}{ds}\cdot\frac{ds}{dt}$$

$$= \frac{d^2s}{dt^2}\ \hat{T}+\left(\frac{ds}{dt}\right)^2\frac{d\hat{T}}{ds}$$

$$= \frac{d^2s}{dt^2}\ \hat{T}+\left(\frac{ds}{dt}\right)^2 k\ \hat{N} \tag{iii}$$

$$\therefore\ \dot{\vec{r}}\times\ddot{\vec{r}} = \left[\frac{ds}{dt}\ \hat{T}\right]\times\left[\frac{d^2s}{dt^2}\ \hat{T}+\left(\frac{ds}{dt}\right)^2 k\ \hat{N}\right]$$

$$= \left(\frac{ds}{dt}\right)^3 k\ \left(\hat{T}\times\hat{N}\right) \qquad \left(\because\hat{T}\times\hat{T}=0\right)$$

$$= \left(\frac{ds}{dt}\right)^3 k\ \hat{B} \tag{iv}$$

$\therefore \left|\dot{\vec{r}} \times \ddot{\vec{r}}\right| = k\left(\frac{ds}{dt}\right)^3$ and $\left|\dot{\vec{r}}\right| = \frac{ds}{dt}$ (From (ii))

Equating the magnitudes, we get

$$\left|\dot{\vec{r}} \times \ddot{\vec{r}}\right| = k\left|\dot{\vec{r}}\right|^3 \tag{v}$$

$$\therefore k = \left|\frac{\dot{\vec{r}} \times \ddot{\vec{r}}}{\left|\dot{\vec{r}}\right|^3}\right| \tag{3.13}$$

Formula (3.13) determines the curvature k or radius of curvature $\varrho\left(=\frac{1}{k}\right)$ at a point P on the curve.

From equation (iii), we get

$$\dddot{\vec{r}} = \frac{d^3\vec{r}}{dt^3} = \frac{d^3s}{dt^3}\hat{T} + \frac{d^2s}{dt^2}\frac{d\hat{T}}{dt} + \frac{d}{dt}\left[k\left(\frac{ds}{dt}\right)^2\right]\hat{N} + k\left(\frac{ds}{dt}\right)^2\frac{d\hat{N}}{dt}$$

$$= \frac{d^3s}{dt^3}\hat{T} + \frac{d^2s}{dt^2}\frac{d\hat{T}}{ds}\cdot\frac{ds}{dt} + \frac{d}{dt}\left[k\left(\frac{ds}{dt}\right)^2\right]\hat{N} + k\left(\frac{ds}{dt}\right)^2\left[\frac{d\hat{N}}{ds}\cdot\frac{ds}{dt}\right]$$

Substituting $\frac{d\hat{T}}{ds}$ and $\frac{d\hat{N}}{ds}$ from the results of the Serret-Frenet Formula, we have

$$\dddot{\vec{r}} = \frac{d^3s}{dt^3}\hat{T} + \frac{d^2s}{dt^2}\frac{ds}{dt}k\hat{N} + \frac{d}{dt}\left[k\left(\frac{ds}{dt}\right)^2\right]\hat{N} + k\left(\frac{ds}{dt}\right)^3\left[\tau\hat{B} - k\hat{T}\right]$$

$$= \left[\frac{d^2s}{dt^3} - k^2\left(\frac{ds}{dt}\right)^3\right]\hat{T} + \left\{k\frac{ds}{dt}\frac{d^2s}{dt^2} + \frac{d}{dt}\left[k\left(\frac{ds}{dt}\right)^2\right]\right\}\hat{N}$$

$$+ \left[k\,\tau\left(\frac{ds}{dt}\right)^3\right]\hat{B}$$

$$\therefore \dddot{\vec{r}}\cdot\left(\dot{\vec{r}} \times \ddot{\vec{r}}\right) = \left[\left(\frac{ds}{dt}\right)^3 k\,\hat{B}\right].\ \dddot{\vec{r}}\ (\because \text{From (iv)})$$

$$= \left(\frac{ds}{dt}\right)^6 k^2\tau\ \left(\because \hat{B} \times \hat{T} = \hat{B} \times \hat{N} = 0\right)$$

$$= \left[\dot{\vec{r}} \times \ddot{\vec{r}}\right]^2\tau\ (\because \text{From (v)})$$

$$\therefore \ \tau = \frac{\dddot{\vec{r}} \cdot \left(\dot{\vec{r}} \times \ddot{\vec{r}}\right)}{\left[\dot{\vec{r}} \times \ddot{\vec{r}}\right]^2} \tag{3.14}$$

The formula (3.14) gives the torsion τ or radius of torsion $\sigma = \left(\frac{1}{\tau}\right)$ at a point P on the curve.

Illustration 3.8: For the curve $x = t, \ y = t^2, \ z = \frac{2}{3}$, find (i) K and (ii) τ at point $'t'$.

Solution:

$$\vec{r} = t\,(i)\,\hat{\imath} + t^2\hat{\jmath} + \left(\frac{2}{3}t^3\right)\hat{k}$$

$$\dot{\vec{r}} = \hat{\imath} + 2t\hat{\jmath} + 2t^2\hat{k}$$

$$\ddot{\vec{r}} = 2\hat{\jmath} + 4t\,\hat{k}$$

$$\dddot{\vec{r}} = 4\hat{k}$$

$$\dot{\vec{r}} \times \ddot{\vec{r}} = \begin{bmatrix} \hat{\imath} & \hat{\jmath} & \hat{k} \\ 1 & 2t & 2t^2 \\ 0 & 2 & 4t \end{bmatrix} = (4t^2)\,\hat{\imath} - (4t)\,\hat{\jmath} + 2\hat{k}$$

Hence, $\left[\vec{r}\right]^2 = 1 + 4t^2 + 4t^4 = \left(1 + 2t^2\right)^2$

and $\left[\dot{\vec{r}} \times \ddot{\vec{r}}\right]^2 = 16t^2 + 16t^2 + 4 = 4\left(2t^2 + 1\right)^2$

and $\left(\dot{\vec{r}} \times \ddot{\vec{r}}\right) \cdot \dddot{\vec{r}} = \left[4t^2\,i - 4t\,j + 2kj\right] \ . \ \left[4k\right] = 8$

Using formula $k = \dfrac{\left[\dot{\vec{r}} \times \ddot{\vec{r}}\right]}{\left[\dot{\vec{r}}\right]^3}$ and $\tau = \dfrac{\left(\dot{\vec{r}} \times \ddot{\vec{r}}\right) \cdot \left(\dddot{\vec{r}}\right)}{\left[\dot{\vec{r}} \times \ddot{\vec{r}}\right]^2}$

we get

$k = \frac{2\left(1+2t^2\right)}{\left(1+2t^2\right)^3} = \frac{2}{\left(1+2t^2\right)^2}$ and $\tau = \frac{8}{4\left(1+2t^2\right)^2} = \frac{2}{\left(1+2t^2\right)^2}$

Illustration 3.9: Show that for the curve $x = a\cos\theta$, $y = a\sin\theta$, $z = a\,\theta\cot\beta$ curvature and torsion are $k = \frac{1}{a}\sin^2\beta$ and $\tau = \frac{1}{a}\sin\beta\cos\beta$.

Solution:

Here, $\vec{r} = (a\cos\theta\ )\,\hat{\imath} + (a\sin\theta\ )\ \hat{\jmath} + (a\,\theta\cot\beta\ )\ \hat{k}$

$$\therefore \ \dot{\vec{r}} \times \ddot{\vec{r}} = \begin{bmatrix} \hat{\imath} & \hat{\jmath} & \hat{k} \\ -a\sin\theta & a\cos\theta & a\cot\beta \\ -a\cos\theta & -a\sin\theta & 0 \end{bmatrix}$$

and

$$\dddot{\vec{r}} = (a \sin \theta\ )\hat{\imath} + (-a \cos \theta\ )\hat{\jmath} + 0\,\hat{k}$$

$$\left[\dot{\vec{r}}\right]^2 = a^2 \sin \theta\ + a^2 \cos^2 \theta + a^2 \cot^2 \beta$$

$$= a^2 \left[1 + \cot^2 \beta\ \right] = a^2\ cosec^2\ \beta$$

$$\left[\dot{\vec{r}} \times \ddot{\vec{r}}\right]^2 = a^4 \sin^2 \theta \cot^2 \beta\ \ + a^2 \cos^2 \theta \cot^2 \beta + a^4 \qquad \left(\cot^2 \beta\ + 1\right)$$

$$= a^4\ cosec^2\ \beta$$

and

$$\left(\dot{\vec{r}} \times \ddot{\vec{r}}\right) \cdot \dddot{\vec{r}} = \left[\left(a^2 \sin \theta \cot \beta\ \right)\hat{\imath} + \left(-a^2 \cos \theta \cot \beta\ \right)\hat{\jmath} + a^2\hat{k}\right] \cdot$$

$$[(a \sin \theta\ )\hat{\imath} + (-a \cos \theta\ )\hat{\jmath}]$$

$$= a^3 \sin^2 \theta\ \ \cot \beta\ + a^3 \cos^2 \theta\ \ \cot \beta\ \ = a^3 \cot \beta$$

$$\therefore k = \frac{\left[\dot{\vec{r}} \times \ddot{\vec{r}}\right]}{\left[\dot{\vec{r}}\right]^3} = \frac{a^2\ cosec\ \beta}{a^3\ cosec^3 \beta} = \frac{1}{a} \sin^2 \beta$$

and

$$\tau = \frac{\left(\dot{\vec{r}} \times \ddot{\vec{r}}\right) \cdot \dddot{\vec{r}}}{\left[\dot{\vec{r}} \times \ddot{\vec{r}}\right]^2} = \frac{a^3 \cot \beta}{a^4\ cosec^2\ \beta} = \frac{1}{a} \sin \beta\ \ \cos \beta$$

3.5 Vector Differentiation

Next, we show the basic concepts of calculus, such as vector differentiation, differentiability defined for scalar, and vector functions in a simple and natural way.

The derivative of a vector function $\vec{v}(t)$ with respect to the scalar, variable t is given as

$$\frac{d\vec{v}}{dt} = \vec{v}'\,(t) = \lim_{\Delta t \to 0} \frac{\vec{v}\,(t + \Delta t) - \vec{v}\,(t)}{\Delta t} \tag{3.15}$$

provided the limit exists.

The vector $\vec{v}'(t)$ is called the derivative of $\vec{v}'(t)$. (see Figure 3.5).

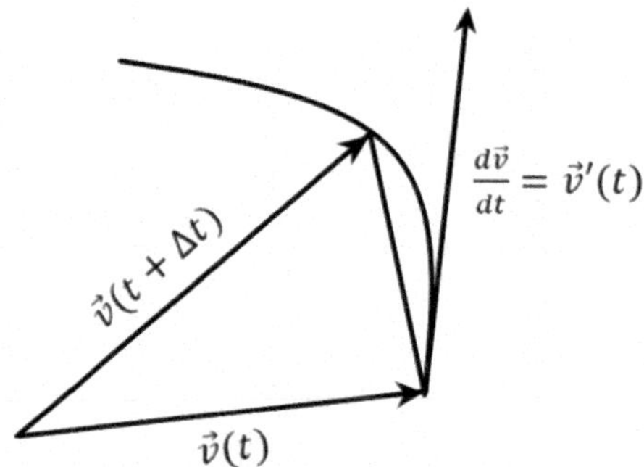

Figure 3.5 Represents the derivative of the vector $\vec{v}(t)$

In terms of components with respect to a given Cartesian coordinate system, $\overrightarrow{v}(t)$ is differentiable at a point t if and only if its three components $v_1(t)$, $v_2(t)$, and $v_3(t)$ are differentiable at t, and then the derivative $\overrightarrow{v}'(t)$ is obtained by differentiating each component separately,

$$\overrightarrow{v}'(t) = \left(v_1',\ v_2',\ v_3'\right) \tag{3.16}$$

Most of the familiar rules of differentiation yield corresponding rules for differentiating vector functions provided the order of factors in vector product is maintained.

1. The derivative of a constant vector is zero.
2. $(c\overrightarrow{v})' = c\,\overrightarrow{v}'$ where c is constant
3. $(\overrightarrow{u} \pm \overrightarrow{v})' = \overrightarrow{u}' \pm \overrightarrow{v}'$
4. $(\overrightarrow{u} \cdot \overrightarrow{v})' = \overrightarrow{u}' \cdot \overrightarrow{v} + \overrightarrow{u} \cdot \overrightarrow{v}'$
5. $(\overrightarrow{u} \times \overrightarrow{v})' = \overrightarrow{u}' \times \overrightarrow{v} + \overrightarrow{u} \times \overrightarrow{v}$
6. $(\overrightarrow{u} \cdot \overrightarrow{v} \cdot \overrightarrow{w})' = (\overrightarrow{u}' \cdot \overrightarrow{v} \cdot \overrightarrow{w}) + (\overrightarrow{u} \cdot \overrightarrow{v}' \cdot \overrightarrow{w}) + (\overrightarrow{u} \cdot \overrightarrow{v} \cdot \overrightarrow{w}')$

Illustration 3.10: If $\overrightarrow{F}(t)$ has a constant magnitude, then prove that $\frac{d\overrightarrow{F}}{dt}$ is perpendicular to $\overrightarrow{F}(t)$.

Solution:

$\overrightarrow{F}(t)$ has a constant magnitude

$\therefore \left|\overrightarrow{F}(t)\right| = \text{constant}$

$$\therefore \vec{F}(t) \cdot \vec{F}(t) = \left|\vec{F}(t)\right|^2 = \text{constant}$$

$$\therefore \frac{d}{dt}\left(\vec{F} \cdot \vec{F}\right) = 0$$

$$\Rightarrow \vec{F} \cdot \frac{d\vec{F}}{dt} + \frac{d\vec{F}}{dt} \cdot \vec{F} = 0$$

$$\Rightarrow 2\vec{F} \cdot \frac{d\vec{F}}{dt} = 0$$

$$\Rightarrow \vec{F} \cdot \frac{d\vec{F}}{dt} = 0$$

Since, $\vec{F} \cdot \frac{d\vec{F}}{dt} = 0$, $\frac{dF}{dt}$ is perpendicular to $F(t)$. Hence proved.

Illustration 3.10: If $\vec{F}(t)$ has a constant direction, then prove that

$$\vec{F} \times \frac{d\vec{F}}{dt} = 0.$$

Solution:
Let $\left|\vec{F}(t)\right| = \phi(t)$. Let $\vec{G}(t)$ be a unit vector in the direction of $\vec{F}(t)$ so that $\vec{F}(t) = \phi(t)\ \vec{G}(t)$.

$$\therefore \frac{d\vec{F}}{dt} = \phi\frac{d\vec{G}}{dt} + \frac{d\phi}{dt}\vec{G} \qquad \text{(i)}$$

If $\vec{F}(t)$ has constant direction, so has $\vec{G}(t)$. Thus $\vec{G}(t)$ is a constant vector and $\frac{d\vec{G}}{dt} = 0$

From (i).

$$\frac{d\vec{F}}{dt} = \frac{d\phi}{dt}\vec{G}$$

$$\therefore \vec{F} \times \frac{d\vec{F}}{dt} = \phi\vec{G} \times \left(\frac{d\phi}{dt}\vec{G}\right)$$

$$= \phi\left(\vec{G} \times \vec{G}\right)\frac{d\phi}{dt} = 0$$

Illustration 3.11: Find the angle between the tangents to the curve $x = t^2,\ y = 2t,\ z = -t^3$ at the points $t = 1$ and $t = -1$.

Solution:

Let $\vec{r}$ be the position vector of any point (x, y, z) on the curve, then

$$\vec{r}(t) = (x(t), y(t), z(t)) = x(t)\hat{i} + y(t)\hat{i} + z(t)\hat{k}$$
$$\therefore \vec{r} = t^2\hat{i} + 2t\hat{j} - t^2\hat{k}$$
$$\therefore \vec{T} = \frac{d\vec{r}}{dt} = 2t\hat{i} + 2\hat{j} - 3t^2\hat{k}$$

is a vector along the tangent at any point t.

$\therefore \vec{T}_1 = 2\hat{i} + 2\hat{j} - 3\hat{k}$ and $\vec{T}_2 = -2\hat{i} + 2\hat{j} - 3\hat{k}$ are the vectors along the tangents at $t = 1$ and $t = -1$ respectively.

If θ be the angle between $\vec{T}_1$ and $\vec{T}_2$, Then

$$\cos\theta = \frac{\vec{T}_1 \cdot \vec{T}_2}{\left|\vec{T}_1\right| \left|\vec{T}_2\right|} = \frac{2(-2) + 2(2) - 3(-3)}{\sqrt{4+4+9}\sqrt{4+4+9}} = \frac{9}{17}$$
$$\therefore \theta = \cos^{-1}\left(\frac{9}{17}\right).$$

3.6 Gradient of a Scalar Field and Directional Derivative

The vector differential operator is denoted by ∇ (read as del or nabla) and defined as

$$\nabla = \hat{i}\frac{\partial}{\partial x} + \hat{j}\frac{\partial}{\partial y} + \hat{k}\frac{\partial}{\partial z}$$

and this vector operator has properties analogous to those of ordinary vectors. It is very useful in defining three important quantities gradient, divergence, and curl.

3.6.1 Gradient of a Scalar Field

For a given scalar function $f(x, y, z)$ the gradient of f, is written as grad f or ∇f is the vector function defined by

$$\text{grad } f = \nabla f = \hat{i}\frac{\partial f}{\partial x} + \hat{j}\frac{\partial f}{\partial y} + \hat{k}\frac{\partial f}{\partial z}$$

3.6.1.1 Properties of Gradient

1. The projection of ∇f in any direction is equal to the derivative of f in that direction.

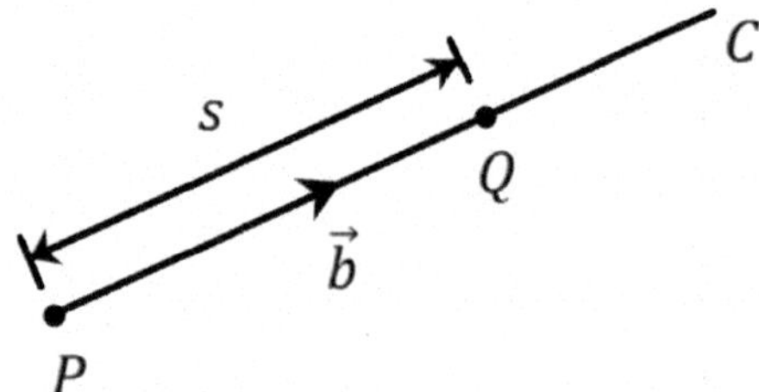

Figure 3.6 Represents the directional derivative

2. The gradient of f is a vector normal to the surface $f(x, y, z) = c =$ constant. So, the angle between any two surfaces $f(x, y, z) = c_1$ and $g(x, y, z) = c_2$ is the angle between their corresponding normal given by ∇f and ∇g respectively.

3. The gradient at any point P gives the maximum rate of change of f in the direction of maximum increase of f at the point P.

3.6.2 Directional Derivative

If $f = f(x, y, z)$ then $\frac{\partial f}{\partial x}, \frac{\partial f}{\partial y}, \frac{\partial f}{\partial z}$ are the derivatives (rate of change) of f in the "direction" of the coordinate axes OX, OY, and OZ respectively. This concept can be extended to define a derivative of f in a given direction $\overrightarrow{PQ}$ (see Figure 3.6)

Let P be a point in space and $\vec{b}$ be a unit vector from P in the given direction. Let s be the arc length measured from P to another point Q along the ray C in the direction $\vec{b}$. Then the directional derivative of $f$at the point P in the direction of $\vec{b}$ is denoted $D_{\vec{b}}f$ or $\frac{df}{ds}$ and defined by (see Figure 3.6)

$$D_{\vec{b}}f = \frac{df}{ds} = \lim_{s \to 0} \frac{f(Q) - f(P)}{s} \quad (s = \text{distance between } P \text{ and } Q)$$

where Q is a variable point on the ray C in the direction of $\vec{b}$ and the ray C is given by

$$\vec{r}(s) = x(s)\hat{i} + y(s)\hat{j} + z(s)\hat{k} \quad (s \geq 0)$$

Next, we use Cartesian coordinates, and for $\vec{b}$ a unit vector. Now consider the function

$$f(s) = f(x, y, z) = f(x(s),\ y(s),\ z(s))$$

and it's derivative with respect to the arc length s of C. Hence, assuming that f has continuous partial derivatives and applying the chain rule, we obtain

$$
\begin{aligned}
D_{\vec{b}} f &= \frac{df}{ds} = \frac{\partial f}{\partial x}\frac{\partial x}{\partial s} + \frac{\partial f}{\partial y}\frac{\partial y}{\partial s} + \frac{\partial f}{\partial z}\frac{\partial z}{\partial s} \\
&= \left(\hat{i}\,\frac{\partial f}{\partial x} + \hat{j}\,\frac{\partial f}{\partial y} + \hat{k}\,\frac{\partial f}{\partial z}\right) \cdot \left(\hat{i}\,\frac{dx}{ds} + \hat{j}\,\frac{dy}{ds} + \hat{k}\,\frac{dz}{ds}\right) \\
&= \left[\left(\hat{i}\,\frac{\partial}{\partial x} + \hat{j}\,\frac{\partial}{\partial y} + \hat{k}\,\frac{\partial}{\partial z}\right) f\right] \cdot \vec{b} \\
&\left(\because \frac{d\vec{r}}{ds} = \hat{i}\,\frac{dx}{ds} + \hat{j}\,\frac{dy}{ds} + \hat{k}\,\frac{dz}{ds}\right) \\
\therefore D_{\vec{b}} f &= \frac{df}{ds} = \nabla f \cdot \vec{b} = \vec{b} \cdot \text{grad } f
\end{aligned}
$$

Thus, the directional derivative of f at any point P is the dot product of unit vector $\vec{b}$ and grad f.

Note that the directional derivative of the vector $\vec{a}$ of any length ($\neq 0$) is

$$D_{\hat{a}} f = \frac{df}{ds} = \text{grad } f \cdot \left(\frac{\vec{a}}{|\vec{a}|}\right) = \nabla f \cdot \frac{1}{|\vec{a}|}\,\vec{a} = \nabla f \cdot \hat{a}$$

where $\hat{a} = \frac{\vec{a}}{|\vec{a}|}$ is the unit vector in the direction of the vector $\vec{a}$.

3.6.2.1 Properties of Gradient

1. The function f increases most rapidly when $\cos\theta = 1$ (i.e., when $\theta = 0$) or when $\vec{a}$ is the direction of ∇f. The derivative in this direction is

$$D_{\vec{a}} f = |\nabla f| \cos(0) = |\nabla f|.$$

2. Similarly, f decreases most rapidly when in the direction of $-\nabla f$. The derivative in this direction is

$$D_{\vec{a}} f = |\nabla f| \cos(\pi) = -|\nabla f|.$$

3. Any direction $\vec{a}$ orthogonal to a gradient $\nabla f \neq 0$ is a direction of zero change in f because θ then equals $\pi/2$ and

$$D_{\vec{a}} f = |\nabla f| \cos\left(\frac{\pi}{2}\right) = |\nabla f| \cdot 0 = 0$$

These properties hold in both two as well as three dimensions.

Illustration 3.12: (a) Gradient as a Surface normal vector, find a unit normal vector n of the cone of revolution $z^2 = 4\left(x^2 + y^2\right)$ at the point P : $(1,\ 0,\ 2)$. (b) If $\emptyset = 3x^2y - y^3z^2$, find grad $\emptyset$ at the point $(1,\ -2,\ -1)$.

Solution:

(a) The cone is a level surface

$f = 0$ for $f\,(x,\ y,\ z) = 4\left(x^2 + y^2\right) - z^2$.

Thus, $\nabla f =$ grand $f = 8x\,\hat{i} + 8y\,\hat{j} - 2z\,\hat{k}$ and at P, grad $f = 8\hat{i} - 4\hat{k}$.

Hence, a unit normal vector of the cone at P is

$$\hat{n} = \frac{1}{|grad\ f|}\ grad\ f = \frac{8\hat{i} - 4\hat{k}}{\sqrt{(8)^2 + (-4)^2}}$$

$$= \frac{8\hat{i} - 4\hat{k}}{\sqrt{80}} = \frac{2}{\sqrt{5}}\,\hat{i} - \frac{1}{\sqrt{5}}\,\hat{k}.$$

(b) Here, $\nabla\phi = 6xy\hat{i} + \left(3x^2 - 3y^2z^2\right)\hat{j} - 2y^3z\hat{k}$

$$\nabla_{at(1,\ -2,\ 1)} = -12\hat{i} - 9\hat{j} - 16\hat{k}.$$

Illustration 3.13: Find the derivative of $f\,(x,\ y) = x^2\text{sin}2y$ at the point $\left(1, \frac{\pi}{2}\right)$ in the direction of $\vec{v} = 3\hat{i} - 4\hat{j}$.

Solution:

Consider,

$$\nabla f = (2x\text{sin}2y\,)\,\hat{i} + \left(2x^2\text{cos}2y\,\right)\hat{j}$$

and

$$(\nabla f)\left(1, \frac{\pi}{2}\right) = 2\text{sin}\pi\ \ \hat{i} + (2\text{cos}\pi\ )\ \hat{j}$$

$$= 0\,\hat{i} - 2\hat{j} = -2\hat{j}$$

The direction of $\vec{v}$ is the unit vector obtained by dividing $\vec{v}$ by its length:

$$\hat{v} = \frac{\vec{v}}{|\vec{v}|} = \frac{3}{5}\,\hat{i} - \frac{4}{5}\,\hat{j}$$

The derivative of at $\left(1, \frac{\pi}{2}\right)$ in the direction of v is therefore

$$\nabla f|_{\left(1,\frac{\pi}{2}\right)} \cdot \hat{v} = \left(-2\hat{j}\right) \cdot \left(\frac{3}{5}\,\hat{i} - \frac{4}{5}\,\hat{j}\right)$$

$$= 0 + \frac{8}{5} = \frac{8}{5}$$

Illustration 3.14: (a) Find the derivative of $f = (x,\ y,\ z) = x^3 - xy^2 - z$ at $P_0\,(1,\ 1,\ 0)$ in the direction of $\vec{v} = 2\hat{i} - 3\hat{j} + 6\hat{k}$. (b) In what directions does f change most rapidly at p_0, and what are the rates of change in these directions?

Solution:

(a) $(\nabla f)\, P_0\,(1,\ 1,\ 0) = 2\hat{i} - 2\hat{j} - \hat{k}$

The direction of $\vec{v}$ is

$$\hat{u} = \frac{\vec{v}}{|\vec{v}|} = \frac{2\hat{i} - 3\hat{j} + 6\hat{k}}{7} = \frac{2}{7}\,\hat{i} - \frac{3}{7}\,\hat{j} + \frac{6}{7}\hat{k}$$

The derivative of f and P_0 in the direction of $\vec{v}$ is

$$\nabla f\,|_{(1,1,0)} \cdot \hat{u} = \left(2\hat{i} - 2\hat{j} - \hat{k}\right) \cdot \left(\frac{2}{7}\,\hat{i} - \frac{3}{7}\,\hat{j} + \frac{6}{7}\,\hat{k}\right)$$

$$= \frac{4}{7} + \frac{6}{7} - \frac{6}{7} = \frac{4}{7}$$

(b) The function increases most rapidly in the direction of $\nabla f = 2\hat{i} - 2\hat{j} - \hat{k}$ and decreases most rapidly in the direction of $-\nabla f$. The rates of change in the directions are, respectively

$$|\nabla f| = \sqrt{(2)^2 + (-2)^2 + (-1)^2} = \sqrt{9} = 3 \text{ and } -|\nabla f| = -3.$$

Illustration 3.15: If $\vec{r} = x\hat{i} + y\hat{j} + z\hat{k}$, Show that

(i) $\nabla r = \hat{r}$

(ii) $\nabla r^n = nr^{n-2}\,\vec{r}$

Solution:

(i) $\text{grad}\, r = \nabla r = \hat{i}\frac{\partial r}{\partial x} + \hat{j}\,\frac{\partial r}{\partial y} + \hat{k}\,\frac{\partial r}{\partial z}$

$$= \hat{i}\left(\frac{x}{r}\right) + \hat{j}\left(\frac{y}{r}\right) + \hat{k}\left(\frac{z}{r}\right)$$

$$= \frac{x\hat{i} + y\hat{j} + z\hat{k}}{r} = \frac{\vec{r}}{r} = \vec{r}$$

$$\left(\because \frac{\vec{r}}{|\vec{r}|} = \hat{r} \text{ unit vector}\right)$$

(ii) $\text{grad}\, r^n = \nabla r^n$

$$= \hat{i}\frac{\partial}{\partial x}\, r^n + \hat{j}\,\frac{\partial}{\partial y}\, r^n + \hat{k}\,\frac{\partial}{\partial z}\, r^n$$

$$= \hat{i}\left(nr^{n-1}\,\frac{\partial r}{\partial x}\right) + \hat{j}\left(nr^{n-1}\,\frac{\partial r}{\partial y}\right) + \hat{k}\left(nr^{n-1}\,\frac{\partial r}{\partial z}\right)$$

$$= \hat{i}\left(nr^{n-1}\,.\,\frac{x}{r}\right) + \hat{j}\left(nr^{n-1}\,.\,\frac{y}{r}\right) + \hat{k}\left(nr^{n-1}\,.\,\frac{z}{r}\right)$$

$$= nr^{n-2}\left(x\hat{i} + y\hat{j} + z\hat{k}\right) = nr^{n-2}\,\vec{r}$$

Illustration 3.16: Find the gradient of
$f\,(x,\ y,\ z) = 2z^3 - 3\left(x^2 + y^2\right) z + \tan^{-1}(xz)$ at $(1,\ 1,\ 1)$.

Solution:

$$\nabla f = \hat{i}\left[-6xz + \frac{1}{1 + x^2z^2}\cdot z\right] + \hat{j}\ [-6yz]$$

$$+\,\hat{k}\left[6z^2 - 3\left(x^2 + y^2\right) + \frac{1}{1 + x^2y^2}\cdot x\right]$$

$$\nabla f_{(1,\ 1,\ 1)} = \hat{i}\left[-6 + \frac{1}{2}\right] + \hat{j}\,[-6] + k\left[6 - 6 + \frac{1}{2}\right]$$

$$= \hat{i}\left[\frac{-11}{2}\right] - 6\hat{j} + \frac{1}{2}\hat{k}$$

$\therefore$ The gradient of f at $(1,\ 1,\ 1) = \frac{1}{2}\left(-11\hat{i} - 12\hat{j} + \hat{k}\right)$

Illustration 3.17: If $\nabla\phi = \left(y^2\right)\hat{i} + \left(2xy + z^3\right)\,\hat{j} + \left(3yz^2\right)\,\hat{k}$, determine ϕ.

Solution:

$$\nabla\phi = \hat{i}\,\frac{\partial\phi}{\partial x} + \hat{j}\,\frac{\partial\phi}{\partial y} + \hat{k}\,\frac{\partial\phi}{\partial z}$$
$$= \hat{i}\,(y^2) + \hat{j}\,(2xy + z^3) + \hat{k}\,(3yz^2)$$
$$\therefore\ \frac{\partial\phi}{\partial x} = y^2 \quad \text{(i)},\ \frac{\partial\phi}{\partial y} = (2xy + z^3) \qquad \text{(ii)},\ \frac{\partial\phi}{\partial z} = 3yz^2 \qquad \text{(iii)}$$

Integrating (i), (ii) and (iii) partially with respect to $x,\ y,\ z$ respectively,

$$\phi = xy^2 + f_1\ (y,\ z)$$
$$\phi = xy^2 + yz^3 + f_2\,(z,\ x)$$
$$\phi = yz^2 + f_1\,(x,\ y)$$
$$\phi = xy^2 + yz^3 + c\ (\text{c is an arbitrary constant})$$

Illustration 3.18: Find the magnitude and the direction of the greatest change of $u = xyz^2$ at $(1,\ 0,\ 3)$.

Solution:

$$\nabla u = \hat{i}\,\frac{\partial u}{\partial x} + \hat{j}\,\frac{\partial u}{\partial y} + \hat{k}\,\frac{\partial u}{\partial z}$$
$$= \hat{i}\ (yz^2) + \hat{j}\ (xz^2) + \hat{k}\ (2xyz)$$

$\therefore\ \nabla u$ at $(1,\ 0,\ 3) = 9\hat{j}$

$\therefore$ The magnitude of the greatest change of $u = |\nabla u| = 9$ and its direction is along the y-axis.

Illustration 3.19: The temperature at any point in space is given by $T = xy + yz + zx$. Determine the derivative of T in the direction of the vector $3\hat{i} - 4\hat{k}$ at the point $(1,\ 1,\ 1)$.

Solution:
The derivative of T in the direction of the vector $3\hat{i} - 4\hat{k}$ at the point $P\,(1,\ 1,\ 1)$ is given by $(\nabla T)_p \cdot \vec{a}$; where $\vec{a} = 3\hat{i} - 4\hat{k}$.

Now,

$$\nabla T = \hat{i}\frac{\partial}{\partial x}\ (xy + yz + zx) + \hat{j}\frac{\partial}{\partial y}\ (xy + yz + zx)$$
$$+ \hat{k}\frac{\partial}{\partial z}\ (xy + yz + zx)$$

$= (y + z)\ \hat{\imath} + (z + x)\ \hat{\jmath} + (x + y)\ \hat{k}$

$\therefore (\nabla T)_{p(1,\ 1,\ 1)} = 2\hat{\imath} + 2\hat{\jmath} + 2\hat{k}$

$\therefore$ Directional Derivative $of\ T = (\nabla T)_P \cdot \hat{a} = \left(2\hat{\imath} + 2\hat{\jmath} + 2\hat{k}\right) \cdot \dfrac{\left(3\hat{\imath} - 4\hat{k}\right)}{\sqrt{9+16}}$

$= \dfrac{2}{5}(6 - 8) = -\dfrac{4}{5}$

Illustration 3.20: Find the directional derivative of $g\,(x,\ y,\ z) = 3e^x \cos(yz)$ at $P_0\,(0,\ 0,\ 0)$ in the direction of $\overrightarrow{A} = 2\hat{\imath} + \hat{\jmath} - 2\hat{k}$.

Solution:

$$\nabla g = \hat{\imath}\frac{\partial g}{\partial x} + \hat{\jmath}\ \frac{\partial g}{\partial y} + \hat{k}\frac{\partial g}{\partial z}$$

$$= \hat{\imath}\ [3\ e^x \cos(yz)\ ] + \hat{\jmath}\ [-3\ e^x \sin(yz)\ \cdot z] + \hat{k}\,[-3\ e^x \sin(yz)\ \cdot y]$$

$$\therefore\ \nabla g \text{ at } p_0\,(0,\ 0,\ 0) = \hat{\imath}\,(3) +\ 0 + 0 = 3\hat{\imath}$$

$$\hat{A} = \frac{2\hat{\imath} + \hat{\jmath} - 2\hat{k}}{\sqrt{4+1+4}} = \frac{2\hat{\imath} + \hat{\jmath} - 2\hat{k}}{3}$$

$$\therefore \text{Directional derivative} = (\nabla g)_{p_0} \cdot \hat{A}$$

$$= 3\hat{\imath}\ \cdot \left(\frac{2\hat{\imath} + \hat{\jmath} - 2\hat{k}}{3}\right)$$

$$= \frac{6 + 0 - 0}{3} = 2$$

Illustration 3.21: Find the derivative of $f\,(x,\ y) = x\ e^y + \cos(xy)$ at the point $(2,\ 0)$ in the direction of $A = 3\hat{\imath} - 4\hat{\jmath}$.

Solution:

$$\hat{A} = \frac{3\hat{\imath} - 4\hat{\jmath}}{\sqrt{9+16}} = \frac{3\hat{\imath} - 4\hat{\jmath}}{5}$$

$\frac{\partial f}{\partial x} = [e^y - y\sin(xy)\ ]$ and $\frac{\partial f}{\partial y} = [x\ e^y -\ x\sin(xy)\ ]$

$\therefore \frac{\partial f}{\partial x}$ at $(2,\ 0) = e^0 - 0 = 1$ and $\frac{\partial f}{\partial y}$ at $(2,\ 0) = 2\,e^0 - 2.0 = 2$

$$\nabla f_{(2,0)} = \hat{\imath}\,(1) + \hat{\jmath}\,(2) = \hat{\imath} + 2\hat{\jmath}$$

The directional derivative of f at $(2,\ 0)$ in the direction $\hat{A}$ is therefore

$$(\nabla f)_{(2,0)} \cdot \hat{A} = \left(\hat{\imath} + 2\hat{\jmath}\right) \cdot \frac{\left(3\hat{\imath} - 4\hat{\jmath}\right)}{5} = \frac{3-8}{5} = \frac{-5}{5} = -1$$

Illustration 3.22: Find the direction in which $f\,(x,\ y) = \frac{x^2}{2} + \frac{y^2}{2}$

(a) increases most rapidly and

(b) decreases most rapidly at the point $(1,\ 1)$.

(c) What are the directions of zero change in f at $(1,\ 1)$?

Solution:

(a) The function increases most rapidly in direction of ∇f at $(1,\ 1)$

The gradient is

$$(\nabla f)_{(1,\ 1)} = \left(x\hat{\imath} + y\hat{\jmath}\right)(1,\ 1) = \hat{\imath} + \hat{\jmath}$$

Its direction is

$$\hat{u} = \frac{\vec{u}}{|\vec{u}|} = \frac{\hat{\imath} + \hat{\jmath}}{\sqrt{(1)^2 + (1)^2}} = \frac{1}{\sqrt{2}}\,\hat{\imath} + \frac{1}{\sqrt{2}}\,\hat{\jmath}$$

(b) The function decreases most rapidly in the direction of $-\nabla f$ at $(1,\ 1)$ which is

$$-\vec{u} = -\frac{1}{\sqrt{2}}\,\hat{\imath} - \frac{1}{\sqrt{2}}\,\hat{\jmath}$$

(c) The directions of zero change at $(1,\ 1)$ are the directions orthogonal to ∇f :

$$\hat{n} = -\frac{1}{\sqrt{2}}\,\hat{\imath} + \frac{1}{\sqrt{2}}\,\hat{\jmath}$$

and

$$-\hat{n} = \frac{1}{\sqrt{2}}\,\hat{\imath} - \frac{1}{\sqrt{2}}\,\hat{\jmath}$$

Since $\hat{u} \cdot \hat{n} = 0$ and $-\hat{u} \cdot (-\hat{n}) = 0$.

Illustration 3.22: Find the directional derivative of $\phi = 4xz^3 - 3x^2y^2z$ at the point $(2, \ -1, \ 2)$

(i) in the direction $2\hat{\imath} + 3\hat{\jmath} + 6\hat{k}$

(ii) towards the point $(1, \ 1, \ -1)$ or vector $\hat{\imath} + \hat{\jmath} - \hat{k}$

(iii) along a line equally inclined with coordinate axes.

(iv) along the tangent to the curve

$x = e^t\text{cost}, \ y = e^t\text{sint}, \quad z = e^t \quad$ at $t = 0$

(v) along the direction normal to the surface

$x^2 + y^2 + z^2 = 9$ at $(1, \ 2, \ 3)$

(vi) along the z-axis.

Solution:

$$\frac{\partial \phi}{\partial x} = 4z^3 - 6xy^2z\,, \ \frac{\partial \phi}{\partial y} = -6x^2yz, \ \frac{\partial \phi}{\partial z} = 12xz^2 - 3x^2y^2$$

$$\therefore \ \nabla\phi = \left(4z^2 - 6xy^2z\right)\hat{\imath} - 6x^2yz\hat{\jmath} + \left(12xz^2 - 3x^2y^2\right)\hat{k}$$

$\nabla\phi$ at $(2, \ -1, \ 2) = 8\hat{\imath} + 48\hat{\jmath} + 84\hat{k}$

(i) $\vec{a} = 2\hat{\imath} + 3\hat{\jmath} + 6\hat{k}$

$$\therefore \hat{a} = \frac{2\hat{\imath} + 3\hat{\jmath} + 6\hat{k}}{\sqrt{4 + 9 + 36}} = \frac{2\hat{\imath} + 3\hat{\jmath} + 6\hat{k}}{7}$$

Directional derivative $= \nabla\phi \cdot \hat{a}$

$$= \left(8\hat{\imath} + 48\hat{\jmath} + 84\hat{k}\right) \cdot \frac{2\hat{\imath} + 3\hat{\jmath} + 6\hat{k}}{7}$$

$$= \frac{16 + 144 + 504}{7} = \frac{664}{7}$$

(ii) an along the line joining say $P\,(2, \ -1, \ 2)$ and $Q \ (1, \ 1, \ -1)$ is

$$\vec{a} = (1 - 2) \ \hat{\imath} + (1 + 1) \ \hat{\jmath} + (-1 - 2) \ \hat{k} = -\hat{\imath} + 2\hat{\jmath} - 3\hat{k}$$

$$\therefore \hat{a} = \frac{-\hat{i} + 2\hat{j} - 3\hat{k}}{\sqrt{14}}$$

Directional derivative $= \nabla\phi \cdot \hat{a}$

$$= \left(8\hat{i} + 48\hat{j} + 84\hat{k}\right) \cdot \frac{\left(-\hat{i} + 2\hat{j} - 3\hat{k}\right)}{\sqrt{14}}$$

$$= \frac{8(-1) + 48(2) + 84(-3)}{\sqrt{14}} = \frac{-164}{\sqrt{14}}$$

(iii) $\overrightarrow{a} = \hat{i} + \hat{j} + \hat{k}$

$$\therefore \hat{a} = \frac{\hat{i} + \hat{j} + \hat{k}}{\sqrt{3}}$$

Directional derivative $= \nabla\phi \,.\hat{a}$

$$= \left(8\hat{i} + 48\hat{j} + 84\hat{k}\right) \cdot \frac{\left(\hat{i} + \hat{j} + \hat{k}\right)}{\sqrt{3}}$$

$$= \frac{140}{\sqrt{3}}$$

(iv) $\frac{dx}{dt} = e^t\,(\text{cos}t - \text{sin}t),\ \frac{dy}{dt} = e^t\,(\text{sin}t + \text{cos}t\,),\ \frac{dz}{dt} = e^t$

$$\therefore \overrightarrow{T} = \frac{d\overrightarrow{r}}{dt} = \frac{dx}{dt}\,\hat{i} + \frac{dy}{dt}\,\hat{j} + \frac{dz}{dt}\,\hat{k}$$

$$\therefore T_{at\ t=0} = \hat{i} + \hat{j} + \hat{k}$$

$$\therefore \hat{T} = \frac{\hat{i} + \hat{j} + \hat{k}}{\sqrt{3}}$$

$\therefore$ Directional derivative $\nabla\phi \cdot \hat{T}$

$$= \left(8\hat{i} + 48\hat{j} + 84\hat{k}\right) \cdot \frac{\left(\hat{i} + \hat{j} + \hat{k}\right)}{\sqrt{3}}$$

$$= \frac{140}{\sqrt{3}}$$

(v) Let $f(x,\ y, z) = x^2 + y^2 + z^2 = c$

$$\therefore\ \nabla f = 2x\hat{\imath} + 2y\hat{\jmath} + 2z\hat{k}$$

Let $\vec{a} = \nabla f_{at(1,\ 2,\ 2)} = 2\hat{\imath} + 4\hat{\jmath} + 4\hat{k}$

$$\therefore \hat{a} = \frac{2\hat{\imath} + 4\hat{\jmath} + 4\hat{k}}{\sqrt{4 + 16 + 16}} = \frac{\hat{\imath} + 2\hat{\jmath} + 2\hat{k}}{3}$$

$\therefore$ Directional derivative $= \nabla\phi \cdot\ \hat{a}$

$$= \left(8\hat{\imath} + 48\hat{\jmath} + 84\hat{k}\right)\ .\ \frac{\hat{\imath} + 2\hat{\jmath} + 2\hat{k}}{3}$$

$$= \frac{8 + 96 + 168}{3} = \frac{272}{3}$$

(vi) Directional derivative $\nabla\phi \cdot \hat{a}\ =\ \nabla\phi \cdot\ \hat{k}$

$$= \left(8\hat{\imath} + 48\hat{\jmath} + 84\hat{k}\right) \cdot \hat{k}$$

$$= 84$$

3.6.3 Equations of Tangent and Normal to the Level Curves

At every point (x_0, y_0) in the domain of $f(x, y)$, the gradient of f is normal to the level curve through (x_0, y_0).

$\therefore$ Equation of the tangent at (x_0, y_0) to the level curve, $f(x, y)$ is

$$(x - x_0)\, f_x\,(x_0, y_0) + (y - y_0)\, f_y\,(x_0, y_0) = 0 \tag{3.17}$$

and the equation of the normal is

$$\frac{(x - x_0)}{f_x(x_0, y_0)} = \frac{(y - y_0)}{f_y(x_0, y_0)} \tag{3.18}$$

Illustration 3.23: Find equations of the tangent and normal to the ellipse $\frac{x^2}{4} + y^2 = 2$ at the point $(-2,\ 1)$.

Solution: The ellipse is a level curve of the function

$$f(x,\ y) = \frac{x^2}{4} + y^2.$$

The gradient f at $(-2,\ 1)$ is

$$\nabla f_{at(-2,\ 1)} = \left(\frac{x}{2}\,\hat{i} + 2y\,\hat{j}\right)_{at(-2,\ 1)} = -\hat{i} + 2\hat{j}$$

The tangent is the line

$$(x+2)\,(-1) + (y-1)\,(2) = 0.$$

or

$$x - 2y = -4$$

Equation of the normal

$$\frac{x+2}{-1} = \frac{y-1}{2}$$

or

$$2x + y = -3$$

3.6.4 Equation of the Tangent Planes and Normal Lines to the Surfaces

The equation of the tangent plane at $P\,(x_0,\ y_0,\ z_0)$ to the surface $z = f\,(x,\ y)$ or $F\,(x,\ y,\ z) = f\,(x,\ y) - z$ is

$$\begin{aligned}&(x-x_0)\ F_x\,(x_0,\ y_0,\ z_0) + (y-y_0)\ F_y\,(x_0,\ y_0,\ z_0)\\&+ (z-z_0)\,F_z\,(x_0,\ y_0,\ z_0) = 0\end{aligned} \tag{3.19}$$

and the equations of the normal lines to the surface through $P\,(x_0,\ y_0,\ z_0\,)$are

$$\frac{x-x_0}{F_x\,(x_0\,,\ y_0,\ z_0)} = \frac{y-y_0}{F_y\,(x_0,\ y_0,\ z_0)} = \frac{z-z_0}{F_z\,(x_0,\ y_0\,,\ z_0)}$$

Illustration 3.24: Find equations for the (a) tangent plane and (b) normal line at the point $P_0\,(1,\ 1,\ 1)$ on the surface $x^2 + y^2 + z^2 = 3$.

Solution:

$$\begin{aligned}&F\,(x,\ y,\ z) = x^2 + y^2 + z^2 - 3\\&\therefore\ \nabla F = 2xi + 2yj + 2zk\\&\therefore\ \nabla F_{atP_0}\,(1,\ 1,\ 1) = 2i + 1j + 2k\end{aligned}$$

$\therefore$ The equation of the tangent plane is

$$2(x-1)+2(y-1)+2(z-1)=0$$

or $2x+2y+2z=6$ or $x+y+z=3$

Normal line

$$\frac{x-1}{2}=\frac{y-1}{2}=\frac{z-1}{2}=t$$

or $x=2t+1,\ y=2t+1,\ z=2t+1$

3.7 Divergence and Curl of a Vector Field

3.7.1 Divergence of a Vector Field

Let $\vec{V}(x,y,z)=V_1\hat{i}+V_2\hat{j}+V_3\hat{k}$ be a differentiable vector function, then the divergence of $\vec{V}$ is denoted by div $\vec{V}$ or $\nabla\cdot\vec{V}$ is the scalar function defined by

$$\text{div}\ \vec{V}=\nabla\cdot\vec{V}=\left(\hat{i}\frac{\partial}{\partial x}+\hat{j}\frac{\partial}{\partial y}+\hat{k}\frac{\partial}{\partial z}\right)\cdot\left(V_1\hat{i}+V_2\hat{j}+V_3\hat{k}\right)$$

$$=\frac{\partial V_1}{\partial x}+\frac{\partial V_2}{\partial y}+\frac{\partial V_3}{\partial z}$$

For example, if $\vec{V}=3xz\hat{i}+2xy\hat{j}-yz^2\hat{k}$, then div $\vec{V}=3z+2x-2yz$.

3.7.1.1 Physical Interpretation of Divergence

Consider the motion of a fluid with velocity $\vec{V}=V_x\hat{i}+V_y\hat{j}+V_z\hat{k}$ at a point $A(x,\ y,\ z)$. Consider a small parallelepiped having one of its corners at $A(x,\ y,\ z)$ and edges parallel to the coordinate axes and having magnitude $\delta x,\ \delta y,\ \delta z$ respectively. (See Figure 3.7)

The mass of the fluid entering through the face $ABCD$ per unit time is $V_y\quad \delta x\ \delta z$ and that flowing out through the opposite face *EFGH* is $V_{y\,+\,\delta y}\quad \delta x\ \delta z=\left(V_y+\frac{\partial V_y}{\partial y}\ \delta y\right)\ \delta x\ \delta z\ by$ using Taylor's series.

Thus, the rate at which fluid flows out from the elementary volume along the y-direction is $\left(V_y+\frac{\partial V_y}{\partial y}\ \delta y\right)\ \delta x\ \delta z-V,\ \delta x\ \delta z=\frac{\partial V_y}{\partial y}\ \delta x\ \delta y\ \delta z.$

Similarly, the rate of outward flow along x and z directions will be given by $\frac{\partial V_x}{\partial x}\ \delta x\ \delta y\ \delta z$ and $\frac{\partial V_z}{\partial z}\ \delta x\ \delta z$ respectively.

Thus, the rate at which fluid flows out of the volume per unit time is $\left(\frac{\partial V_x}{\partial x}+\frac{\partial V_y}{\partial y}+\frac{\partial V_z}{\partial z}\right)\ \delta x\ \delta y\ \delta z.$

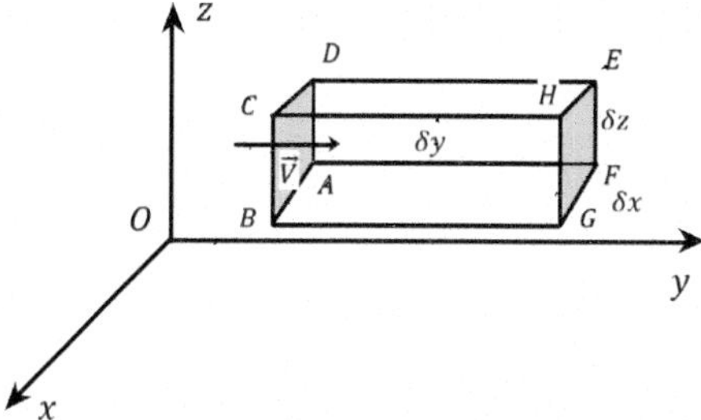

Figure 3.7 Represents the parallelopiped

Dividing this by the volume $\delta x,\ \delta y,\ \delta z$ of the parallelepiped, we have the rate of outflow (or rate of loss of fluid) per unit volume per unit time is

$$\frac{\partial V_x}{\partial x} + \frac{\partial V_y}{\partial y} + \frac{\partial V_z}{\partial z} = div\ \overrightarrow{V}.$$

Hence, the divergence of $\overrightarrow{V}$ gives the rate of outward flow per unit volume at a point of the fluid.

From this, we have the following observations.

(1) If there is no gain of the fluid anywhere, then div $\overrightarrow{V} = 0$ i.e., $\nabla \cdot \overrightarrow{V} = 0$. This is called the continuity equation for an incompressible fluid or condition of incompressibility.

(2) Since the fluid is neither created nor destroyed at any point, it is said to have no sources or sinks.

(3) If the flux entering any element of the space is the same as the leaving it, i.e., if div $\overrightarrow{V} = 0$ everywhere then such a vector point function is called a solenoidal vector function or solenoidal

(4) If $\overrightarrow{V}$ represents an electric flux, div V is the amount of flux that diverges per unit volume.

(5) If $\overrightarrow{V}$ represents heat flux, div $\overrightarrow{V}$ is the rate at which heat flows from a point per unit volume.

Illustration 3.25: If $\overrightarrow{r} = x\hat{i} + y\hat{j} + z\hat{k}$, show that div $r = 3$.

Solution: We know that

$$\text{div}\ \overrightarrow{V} = \nabla \cdot \overrightarrow{V} = \left(\hat{i}\frac{\partial}{\partial x} + \hat{j}\frac{\partial}{\partial y} + \hat{k}\frac{\partial}{\partial z}\right) \cdot \left(V_1\hat{i} + V_2\hat{j} + V_3\hat{k}\right)$$

$$= \frac{\partial V_1}{\partial x} + \frac{\partial V_2}{\partial y} + \frac{\partial V_3}{\partial z}$$

$$\therefore \text{div}\ \overrightarrow{r} = \nabla \cdot \overrightarrow{r} = \tfrac{\partial}{\partial x}\ (x) + \tfrac{\partial}{\partial y}\ (y) + \tfrac{\partial}{\partial z}\ (z)$$

$$= 1 + 1 + 1 = 3$$

Illustration 3.26: If $\overrightarrow{F} = x^2 z\ \hat{i} - 2y^3 z^2\ \hat{j} + xy^2 z\ \hat{k}$ then find div $\overrightarrow{F}$ at the point $(1, -1, 1)$.

Solution: We know that

$$\text{div}\ \overrightarrow{V} = \nabla \cdot \overrightarrow{V} = \left(\hat{i}\frac{\partial}{\partial x} + \hat{j}\frac{\partial}{\partial y} + \hat{k}\frac{\partial}{\partial z}\right) \cdot \left(V_1\hat{i} + V_2\hat{j} + V_3\hat{k}\right)$$

$$= \frac{\partial V_1}{\partial x} + \frac{\partial V_2}{\partial y} + \frac{\partial V_3}{\partial z}$$

$$\therefore \text{div}\ \overrightarrow{F} = \nabla \cdot \overrightarrow{F} = \tfrac{\partial}{\partial x}\ (x^2 z) + \tfrac{\partial}{\partial y}\ (-2y^3 z^2) + \tfrac{\partial}{\partial z}\ (xy^2 z)$$

$$= 2xz - 6y^2 z^2 + xy^2$$

at point $(1, -1, 1)$

$$\therefore\ \nabla \cdot \overrightarrow{F}_{(1,-1,1)} = 2\,(1)\,(1) - 6(-1)^2(1)^2 + (1)\,(-1)^2 = -3.$$

Illustration 3.27: Show that the vector $\overrightarrow{V} = (x + 3y)\,\hat{i} + (y - 2z)\,\hat{j} + (x - 2z)\,\hat{k}$ is solenoidal.

Solution: *A* vector *V* is solenoidal if its divergence is zero.

$$\nabla \cdot \overrightarrow{V} = \frac{\partial}{\partial x}\ (x + 3y) + \frac{\partial}{\partial y}\ (y - 2z) + \frac{\partial}{\partial z}\ (x - 2z)$$

$$= 1 + 1 - 2 = 0.$$

Since $\nabla \cdot \overrightarrow{V} = 0$ therefore vector $\overrightarrow{V}$ is solenoidal.

3.7.2 Curl of a Vector Field

Let $\overrightarrow{V} = (x,\ y,\ z) = V_1\hat{i} + V_2\hat{j} + V_3\hat{k}$ be a differentiable vector function, then the curl or rotation of $\overrightarrow{V}$, written curl $\overrightarrow{V}$ or $\nabla \times \overrightarrow{V}$ is the vector function defined by

$$\text{curl } \overrightarrow{V} = \nabla \times \overrightarrow{V} = \begin{vmatrix} \hat{i} & \hat{j} & \hat{k} \\ \frac{\partial}{\partial x} & \frac{\partial}{\partial y} & \frac{\partial}{\partial z} \\ V_1 & V_2 & V_3 \end{vmatrix}$$

$$= \hat{i}\left(\frac{\partial V_3}{\partial y} - \frac{\partial V_2}{\partial z}\right) + \hat{j}\left(\frac{\partial V_1}{\partial z} - \frac{\partial V_3}{\partial x}\right) + \hat{k}\left(\frac{\partial V_2}{\partial x} - \frac{\partial V_1}{\partial y}\right)$$

3.7.2.1 Physical Interpretation of Curl

To understand the curl of a vector physically, consider the motion of a rigid body rotating about a fixed axis passing through O. Let $\overrightarrow{\omega} = \omega_1\hat{i}+\omega_2\hat{j}+\omega_3\hat{k}$ be the angular velocity of the rigid body, $\overrightarrow{V}$ the linear velocity of any point $P(\overrightarrow{r})$ on the body, where $\overrightarrow{r} = x\hat{i} + y\hat{j} + z\hat{k}$, then

$$\overrightarrow{V} = \overrightarrow{\omega} \times \overrightarrow{r}$$

$$\therefore \overrightarrow{V} = \overrightarrow{\omega} \times \overrightarrow{r} = \begin{vmatrix} \hat{i} & \hat{j} & \hat{k} \\ \omega_1 & \omega_2 & \omega_3 \\ x & y & z \end{vmatrix}$$

$$= \hat{i}\,(\omega_2 z - \omega_3 y) + \hat{j}\,(\omega_3 x - \omega_1 z) + \hat{k}(\omega_1 y - \omega_2 x)$$

Now,

$$\text{curl } \overrightarrow{V} = \nabla \times \overrightarrow{V} = \begin{vmatrix} \hat{i} & \hat{j} & \hat{k} \\ \frac{\partial}{\partial x} & \frac{\partial}{\partial y} & \frac{\partial}{\partial z} \\ \omega_2 z - \omega_3 y & \omega_3 x - \omega_1 z & \omega_1 y - \omega_2 x \end{vmatrix}$$

$$= (\omega_1 + \omega_1)\,\hat{i} + (\omega_2 + \omega_2)\,\hat{j} + (\omega_3 + \omega_3)\,\hat{k}$$

$$= 2\left(\omega_1\hat{i} + \omega_2\hat{j} + \omega_3\hat{k}\right)$$

$$\therefore \text{curl } \overrightarrow{V} = 2\,\overrightarrow{\omega} \Rightarrow \overrightarrow{\omega} = \frac{1}{2}\text{ curl } \overrightarrow{V}$$

Thus, the angular velocity of rotation at any point is equal to half the curl of the linear velocity at that point of the body.

This indicates that the curl of a vector field has something to do with the rotational properties of the field. If the field $\overrightarrow{V}$ is that due to a moving fluid, for example, then a paddlewheel placed at various points in the field would tend to rotate in regions where curl $\overrightarrow{V} \neq 0$, while if curl $\overrightarrow{V}$=0 in the region there would be no rotation and the field $\overrightarrow{V}$ is then called irrotational. A field that is not irrotational is sometimes called a vortex field.

Illustration 3.28: If $\overrightarrow{r} = x\hat{i} + y\hat{i} + z\hat{k}$. show that curl $\overrightarrow{r} = 0$.

Solution:

$$\text{curl } \overrightarrow{r} = \nabla \times \overrightarrow{r} = \begin{vmatrix} \hat{i} & \hat{j} & \hat{k} \\ \frac{\partial}{\partial x} & \frac{\partial}{\partial y} & \frac{\partial}{\partial z} \\ x & y & z \end{vmatrix}$$

$$= \hat{i}\left(\frac{\partial z}{\partial y} - \frac{\partial y}{\partial z}\right) + \hat{j}\left(\frac{\partial}{\partial z}x - \frac{\partial}{\partial x}z\right) + \hat{k}\left(\frac{\partial y}{\partial x} - \frac{\partial x}{\partial y}\right)$$

$$= \hat{i}\,(0) + \hat{j}\,(0) + \hat{k}\,(0) = 0$$

Illustration 3.29: If $\overrightarrow{F} = xz^3\hat{i} - 2x^2yz\hat{j} + 2yz^4\hat{k}$, find curl $\overrightarrow{F}$ at the point $(1,\ 1,\ 1)$.

Solution:

$$\nabla \times \overrightarrow{F} = \begin{vmatrix} \hat{i} & \hat{j} & \hat{k} \\ \frac{\partial}{\partial x} & \frac{\partial}{\partial y} & \frac{\partial}{\partial z} \\ xz^3 & -2x^2yz & 2yz^4 \end{vmatrix}$$

$$\therefore \nabla \times \overrightarrow{F} = \hat{i}\left[\frac{\partial}{\partial y}\left(2yz^4\right) - \frac{\partial}{\partial z}\left(-2x^2yz\right)\right] + \hat{j}\left[\frac{\partial}{\partial z}\left(xz^3\right) - \frac{\partial}{\partial x}\left(2yz^4\right)\right]$$

$$+ \hat{k}\left[\frac{\partial}{\partial z}\left(-2x^2yz\right) - \frac{\partial}{\partial y}\left(xz^3\right)\right]$$

$$= \left(2x^2 + 2x^2y\right)\hat{i} + 3xz^2\hat{j} - 4xyz\hat{k}$$

$$\therefore \text{curl } \overrightarrow{F} \text{ at } (1,\ -1,\ 1) = 3\hat{j} + 4\hat{k}.$$

Illustration 3.30: Find constants $a,\ b,\ c$ so that $\overrightarrow{V} = (x + 2y + az)\,\hat{i} + (bx - 3y - z)\ \hat{j} + (4x + cy + 2z)\,\hat{k}$ is irrotational.

Solution:

$$\text{curl } \overrightarrow{V} = \nabla \times \overrightarrow{V} = \begin{vmatrix} \overrightarrow{i} & \overrightarrow{j} & \overrightarrow{k} \\ \frac{\partial}{\partial x} & \frac{\partial}{\partial y} & \frac{\partial}{\partial z} \\ x + 2y + az & bx - 3y - z & 4x + cy + 2z \end{vmatrix}$$

$$= (c + 1)\,\hat{i} + (a - 4)\ \hat{j} + (b - 2)\,\hat{k}$$

This equals zero when $a = 4$, $b = 2$, $c = -1$ and

$$\vec{V} = (x + 2y + 4z)\ \hat{i} + (2x - 3y - z)\,\hat{j} + (4x - y + 2z)\ \hat{k}$$

Illustration 3.31: If a vector $\vec{F}$ is irrotational then show that $\vec{F} = \nabla\phi =$ grad ϕ, where ϕ is a scalar point function of $\vec{F}$ or scalar potential of $\vec{F}$.

Solution: Let curl $\vec{F} = 0$ where $\vec{F} = F_1\hat{i} + F_2\hat{j} + F_3\hat{k}$

$$\therefore \begin{vmatrix} \hat{i} & \hat{j} & \hat{k} \\ \frac{\partial}{\partial x} & \frac{\partial}{\partial y} & \frac{\partial}{\partial z} \\ F_1 & F_2 & F_3 \end{vmatrix} = 0$$

i.e.,

$$\hat{i}\left(\frac{\partial F_3}{\partial y} - \frac{\partial F_2}{\partial z}\right) + \hat{j}\left(\frac{\partial F_1}{\partial z} - \frac{\partial F_3}{\partial x}\right) + \hat{k}\left(\frac{\partial F_2}{\partial x} - \frac{\partial F_1}{\partial y}\right) = 0$$

$$\therefore \quad \frac{\partial F_3}{\partial y} - \frac{\partial F_2}{\partial z} = 0, \quad \frac{\partial F_1}{\partial z} - \frac{\partial F_3}{\partial x} = 0, \quad \frac{\partial F_2}{\partial x} - \frac{\partial F_1}{\partial y} = 0$$

i.e.,

$$\frac{\partial F_3}{\partial y} = \frac{\partial F_2}{\partial z}, \quad \frac{\partial F_1}{\partial z} = \frac{\partial F_3}{\partial x}, \quad \frac{\partial F_2}{\partial x} = \frac{\partial F_1}{\partial y}$$

These three conditions are satisfied when

$$F_1 = \frac{\partial \phi}{\partial x}, \quad F_2 = \frac{\partial \phi}{\partial y}, \quad F_3 = \frac{\partial \phi}{\partial z},$$

where ϕ is a function of $x,\ y,\ z$

$$\therefore \vec{F} = \hat{i}\ \frac{\partial \phi}{\partial x} + \hat{j}\ \frac{\partial \phi}{\partial y} + \hat{k}\ \frac{\partial \phi}{\partial z} = \nabla\phi$$

ϕ is known as the scalar point function of F. A vector field $\vec{F}$ which can be derived from a scalar field ϕ so that $\vec{F} = \nabla\phi$ is called a conservative vector field and ϕ is called the scalar potential.

Note that conversely if $\vec{F} = \nabla\phi$, then $\nabla \times \vec{F} = 0$.

Illustration 3.32: Prove that $\vec{F} = (y^2 \cos x + z^3)\ \hat{i} + (2y \sin x - 4)\hat{j} + 3xz^2\ \hat{k}$ is irrotational and find its scalar potential.

Solution:

$$\nabla \times \vec{F} = \begin{vmatrix} \hat{i} & \hat{j} & \hat{k} \\ \frac{\partial}{\partial x} & \frac{\partial}{\partial y} & \frac{\partial}{\partial z} \\ y^2\text{cos}x + z^3 & 2y\text{sin}x - 4 & 3xz^2 \end{vmatrix} = 0$$

on simplification

Hence,

$$\vec{F} = \nabla\phi = \hat{i}\,\frac{\partial\phi}{\partial x} + \hat{j}\,\frac{\partial\phi}{\partial y} + \hat{k}\,\frac{\partial\phi}{\partial z}$$

$$\therefore \text{(i)}\ \frac{\partial\phi}{\partial x} = y^2\text{cos}x + z^3\ ,\ \text{(ii)}\frac{\partial\phi}{\partial y} = 2y\text{sin}x - 4\ ,\ \text{(iii)}\frac{\partial\phi}{\partial z} = 3xz^2$$

$\therefore$ from (i), $\phi = y^2\text{sin}x + xz^3 + f_1(y,\ z)$.

from (ii), $\phi = y^2\text{sin}x - 4y + f_2(z,\ x)$.

from (iii), $\phi = xz^3 + f_3(x,\ y)$.

$\therefore \phi = y^2\text{sin}x + xz^3 - 4y + c$, where c is an arbitrary constant.

Illustration 3.33: A fluid motion is given by

$$\vec{V} = (y\text{sin}z\ - \text{sin}x\ )\ \hat{i} + (x\text{sin}z\ + 2yz)\,\hat{j} + \left(xy\text{cos}z\ + y^2\right)\hat{k}.$$

is motion irrotational? If so, find the velocity potential.

Solution:

$$\begin{aligned} \text{Curl}\ \vec{V} &= \begin{vmatrix} \hat{i} & \hat{j} & \hat{k} \\ \frac{\partial}{\partial x} & \frac{\partial}{\partial y} & \frac{\partial}{\partial z} \\ ysinz - sinx & xsinz + 2yz & xycosz + y^2 \end{vmatrix} \\ &= \hat{i}\,[x\ cosz + 2y - x\ cosz - 2y] + \hat{j}\ [\,y\ cosz - y\ cosz] \\ &\quad + \hat{k}\ [sinz - sinz] \\ &= \hat{i}\,(0) + \hat{j}\,(0) + \hat{k}\,(0) = 0 \end{aligned}$$

Since curl $\vec{V} = 0$, the motion is irrotational.

Hence $\vec{V} = \nabla\phi = \hat{i}\,\frac{\partial\phi}{\partial x} + \hat{j}\,\frac{\partial\phi}{\partial y} + \hat{k}\frac{\partial\phi}{\partial z}$

$\frac{\partial\phi}{\partial x} = y\ sinz - sinx$ (i)

$\frac{\partial\phi}{\partial y} = y\ sinz + 2yz$ (ii)

$\frac{\partial\phi}{\partial z} = xy\ cosz + y^2$ (iii)

From (i), $\phi = xy\ sinz + cosx + f_1(y,\ z)$

From (ii), $\phi = xy\ sinz + y^2z + f_2\,(z,\ x)$

From (iii), $\phi = xy\ sinz + zy^2 + f_3\,(x,\ y)$

$\therefore \phi = xy\ sinz + y^2z + cosx + c$, where c an arbitrary constant is required velocity potential.

Illustration 3.34: If $\overrightarrow{A} = 2yz\hat{\imath} - x^2y\hat{\jmath} + xz^2\hat{k},\ \overrightarrow{B} = x^2\hat{\imath} + yz\hat{\jmath} - xy\hat{k}$, and $\phi = 2x^2yz^3$, find

(i) $\left(\overrightarrow{A} \cdot \nabla\right)\phi$

(ii) $\overrightarrow{A} \cdot\ \nabla\phi$ and compare them.

(iii) $\left(\overrightarrow{A} \times \nabla\right)\phi$

(iv) $\overrightarrow{A} \times \nabla\phi$ and compare them.

Solution:

(i)

$$\begin{aligned}
(A \cdot \nabla)\ \phi &= \left[\left(2yz\hat{\imath} - x^2y\hat{\jmath} + xz^2\hat{k}\right) \cdot \left(\hat{\imath}\,\frac{\partial}{\partial x} + \hat{\jmath}\,\frac{\partial}{\partial y} + \hat{k}\,\frac{\partial}{\partial z}\right)\right]\phi \\
&= \left(2yz\frac{\partial}{\partial x} - x^2y\frac{\partial}{\partial y} + xz^2\frac{\partial}{\partial z}\right)\ (2x^2yz^3) \\
&= 2yz\frac{\partial}{\partial x}\ (2x^2yz^3) - x^2y\frac{\partial}{\partial y}\ (2x^2yz^3) + xz^2\ \frac{\partial}{\partial z}\ (2x^2yz^3) \\
&= (2yz)\,(4xyz^3) - (x^2y)\,(2x^2z^3) + (xz^2)\,(6x^2yz^2) \\
&= 8xy^2z^4 - 2x^4yz^3 + 6x^3yz^4
\end{aligned}$$

(ii)

$$\begin{aligned}
\overrightarrow{A} \cdot \nabla\phi &= \left(2yz\hat{\imath} - x^2y\hat{\jmath} + xz^2\hat{k}\right) \cdot \left(\frac{\partial\phi}{\partial x}\,\hat{\imath} + \frac{\partial\phi}{\partial y}\,\hat{\jmath} + \frac{\partial\phi}{\partial z}\,\hat{k}\right) \\
&= \left(2yz\hat{\imath} - x^2y\hat{\jmath} + xz^2\hat{k}\right) \cdot \left(4xyz^3\hat{\imath} + 2x^2z^3\hat{\jmath} + 6x^2yz^2\hat{k}\right) \\
&= 8xy^2z^4 - 2x^4yz^3 + 6x^3yz^4
\end{aligned}$$

Comparison with (i) illustrates the result $\left(\overrightarrow{A} \cdot \nabla\right)\phi = A \cdot \nabla\phi$.

(iii)

$$(\vec{A} \times \nabla)\phi = \left[\left(2yz\hat{i} - x^2y\hat{j} + xz^2\hat{k}\right) \times \left(\hat{i}\frac{\partial}{\partial x} + \hat{j}\,\frac{\partial}{\partial y} + \hat{k}\,\frac{\partial}{\partial z}\right)\right]\,\phi$$

$$= \begin{vmatrix} \hat{i} & \hat{j} & \hat{k} \\ 2yz & -x^2y & xz^2 \\ \frac{\partial}{\partial x} & \frac{\partial}{\partial y} & \frac{\partial}{\partial z} \end{vmatrix} \phi$$

$$= \left[\hat{i}\left(-x^2y\frac{\partial}{\partial z} - xz^2\,\frac{\partial}{\partial y}\right) + \hat{j}\left(xz^2\,\frac{\partial}{\partial x} - 2yz\frac{\partial}{\partial z}\right)\right.$$

$$\left. + \hat{k}\left(2yz\frac{\partial}{\partial y} + x^2y\frac{\partial}{\partial x}\right)\right]\phi$$

$$= -\left(x^2y\frac{\partial\phi}{\partial z} + xz^2\frac{\partial\phi}{\partial y}\right)\,\hat{i} + \left(xz^2\frac{\partial\phi}{\partial x} - 2yz\frac{\partial\phi}{\partial z}\right)\,\hat{j}$$

$$+ \left(2yz\frac{\partial\phi}{\partial y} - x^2y\frac{\partial\phi}{\partial x}\right)\hat{k}$$

$$\left(\vec{A} \times \nabla\right)\,\phi = -\left(6x^4y^2z^2 + 2x^3z^5\right)\,\hat{i} + \left(4x^2yz^5\right)\,\hat{i}$$

$$+ \left(4x^2yz^5 - 12x^2y^2z^3\right)\,\hat{j} + \left(4x^2yz^4 + 4x^3y^2z^3\right)\,\hat{k}$$

(iv)

$$\left(\vec{A} \times \nabla\right)\phi = \left(2yz\hat{i} - x^2y\hat{j} + xz^2\hat{k}\right) \times \left(\hat{i}\,\frac{\partial\phi}{\partial x} + \hat{j}\,\frac{\partial\phi}{\partial y} + \hat{k}\frac{\partial\phi}{\partial z}\right)$$

$$= \begin{vmatrix} \hat{i} & \hat{j} & \hat{k} \\ 2yz & -x^2y & xz^2 \\ \frac{\partial\phi}{\partial x} & \frac{\partial\phi}{\partial y} & \frac{\partial\phi}{\partial z} \end{vmatrix}$$

$$= \left(-x^2\,y\frac{\partial\phi}{\partial z} - xz^2\frac{\partial\phi}{\partial y}\right)\,\hat{i} + \left(xz^2\,\frac{\partial\phi}{\partial x} - 2yz\frac{\partial\phi}{\partial z}\right)\,\hat{j}$$

$$+ \left(2yz\frac{\partial\phi}{\partial y} + x^2y\frac{\partial\phi}{\partial x}\right)\,\hat{k}$$

$$= -\left(6x^4y^2z^2 + 2x^3z^5\right)\,\hat{i} + \left(4x^2yz^5 - 12x^2y^2z^3\right)\,\hat{j}$$

$$+ \left(4x^2y^3z^4 + 4x^3y^2z^3\right)\,\hat{k}$$

Comparison with (iii) illustrates the result $\left(\vec{A} \times \nabla\right)\,\phi = \vec{A} \times \nabla\phi$.

Illustration 3.35: Find the scalar potential function f for

$$\overrightarrow{A} = y^2\hat{\imath} + 2xy\hat{\jmath} - z^2\hat{k}\,.$$

Solution:

$$\text{Curl}\ \overrightarrow{A} = \nabla \times \overrightarrow{A} = \begin{vmatrix} \hat{\imath} & \hat{\jmath} & \hat{k} \\ \frac{\partial}{\partial x} & \frac{\partial}{\partial y} & \frac{\partial}{\partial z} \\ y^2 & 2xy & -z^2 \end{vmatrix}$$

$$= \hat{\imath}\,(0) + \hat{\jmath}\ (0) + \hat{k}\,(2y - 2y) = 0$$

Since and $\overrightarrow{A} = 0$, $\overrightarrow{A} = \nabla\phi$, where ϕ is a scalar potential function of $\overrightarrow{A}$

$$\overrightarrow{A} = \nabla\phi \Longrightarrow \left(y^2\hat{\imath} + 2xy\hat{\jmath} - z^2\hat{\jmath}\right) = \hat{\imath}\,\frac{\partial\phi}{\partial x} + \hat{\jmath}\frac{\partial\phi}{\partial y} + \hat{k}\frac{\partial\phi}{\partial z}$$

$\frac{\partial\phi}{\partial x} = y^2$ (i)
$\frac{\partial\phi}{\partial y} = 2xy$ (ii)
$\frac{\partial\phi}{\partial z} = -z^2$ (iii)
from (i) $\phi = xy^2 + f_1\,(y,\ z)$
from (ii) $\phi = 2xy^{2/2} + f_2\,(z,\ x) = xy^2 + f_2\,(y, z)$
from (iii) $\phi = -2^3/3 + f_3\,(x,\ y)$
$\therefore\ \phi = xy^2 - z^3/3 + C.$ where C is an arbitrary constant.

Illustration 3.36: Prove that $r^n\ \overrightarrow{r}$ is irrotational.

Solution:

To prove $r^n\ \overrightarrow{r}$ irrotational, find $\nabla \times (r^n\ \overrightarrow{r})$ i.e., curl $(r^n\ \overrightarrow{r})$.

$$curl\,(r^n\overrightarrow{r}) = \sum \hat{\imath} \times \frac{\partial}{\partial x}\,(r^n\ ,\ \overrightarrow{r})$$

$$= \sum \hat{\imath}\ \times \left\{nr^{n-1}\,\hat{r}\frac{x}{r} + r^n\ \hat{\imath}\right\}$$

$$= \sum \{nr^{n-2} \times \left(\hat{\imath} \times \overrightarrow{r}\right) + r^n(\hat{\imath} \times \hat{\imath})\}$$

$$= \sum n\ r^{n-2}\ x\left(y\hat{k} - z\hat{\jmath}\right) + 0$$

$$= n\ r^{n-2}\left\{x\ \left(y\hat{k} - z\hat{\jmath}\right) + \ y\left(z\hat{\imath} - x\hat{k}\right) + z\left(x\hat{\jmath} - y\hat{\imath}\right)\right\}$$

$$= 0$$

$\therefore r^n\ \overrightarrow{r}$ is irrotational.

Illustration 3.37: Show that $\overrightarrow{F} = 2xy\ z\hat{i} + \left(x^2z + 2y\right)\ \hat{j} + x^2y\ \hat{k}$ is irrotational and find a scalar function ϕ such that $\overrightarrow{F} =$ grad $\emptyset$.

Solution:

$$\nabla \times \overrightarrow{F} = \begin{vmatrix} \hat{i} & \hat{j} & \hat{k} \\ \frac{\partial}{\partial x} & \frac{\partial}{\partial y} & \frac{\partial}{\partial z} \\ 2xyz & x^2z + 2y & x^2y \end{vmatrix}$$

$$= \hat{i}\ \left(x^2 - x^2\right) + \hat{j}\ (2xy - 2xy) + \hat{k}\ (2xz - 2xz) = 0$$

$\therefore \overrightarrow{F}$ is irrotational

$F =$ grad $\phi \implies$ (i) $\frac{\partial \phi}{\partial x} = 2xyz$, (ii) $\frac{\partial \phi}{\partial y} = x^2z + 2y$, (iii) $\frac{\partial \phi}{\partial z} = x^2y$

Integrating (i), (ii) and (iii) w.r.t. $x,\ y$, and z respectively, we get

$$\phi = x^2yz + f_1\ (y,\ z)$$
$$\phi = x^2yz + y^2 + f\ (z,\ x)$$
$$\phi = x^2yz + y^2 + f_3\ (x,\ y)$$
$$\therefore \phi = x^2yz + y^2 + c$$

3.7.3 Formulae for grad, div, curl Involving Operator ∇

3.7.3.1 Formulae for grad, div, curl Involving Operator ∇ Once

1. If $\overrightarrow{u}$ and $\overrightarrow{v}$ are vector point functions and f, a scalar point function.

   (i) div $\left(\overrightarrow{u} + \overrightarrow{v}\right) =$ div $\overrightarrow{u} +$ div $\overrightarrow{v}$
   or $\nabla \cdot \left(\overrightarrow{u} + \overrightarrow{v}\right) = \nabla \cdot \overrightarrow{u} + \nabla \cdot \overrightarrow{v}$

   (ii) div $\left(f\overrightarrow{u}\right) = f$ (div $\overrightarrow{u}$) + (grad f)$\cdot\ \overrightarrow{u}$
   or $f\ \left(\nabla \cdot \overrightarrow{u}\right) + (\nabla f) \cdot \overrightarrow{u}$.

2. (i) $\nabla \times \left(f\overrightarrow{u}\right) = (\nabla f) \times \overrightarrow{u} + f\ \left(\nabla \times \overrightarrow{u}\right)$
   or curl $\left(f\overrightarrow{u}\right) =$ (grad f) $\times \overrightarrow{u} + f$ curl $\overrightarrow{u}$

   (ii) $\nabla \times \left(\overrightarrow{u} \times \overrightarrow{v}\right) = \left(\nabla \times \overrightarrow{v}\right) \overrightarrow{u} - \left(\nabla \cdot \overrightarrow{u}\right)\ \overrightarrow{v} + \left(\overrightarrow{v} \cdot \nabla\right)\ \overrightarrow{u} - \left(\overrightarrow{u} \cdot \nabla\right)\ \overrightarrow{v}$

3. If u and v are scalar point functions

   (i) grad $(u + v) =$ grad $u+$ grad v

   (ii) grad $(uv) = u$ (grad v) + (grad u) v

   The proofs for (3) are very simple so left to the reader.

Proof:

1. (i) $\nabla \cdot (\vec{u} + \vec{v}) = \left(\hat{i}\frac{\partial}{\partial x} + \hat{j}\,\frac{\partial}{\partial y} + \hat{k}\frac{\partial}{\partial z}\right) \cdot (\vec{u} + \vec{v})$

$$= \sum \hat{i}\,\frac{\partial}{\partial x}\,\cdot (\vec{u} + \vec{v})$$

$$= \sum \hat{i}\,\frac{\partial}{\partial x}\cdot\,\vec{u} + \sum \hat{i}\,\frac{\partial}{\partial x}\cdot\,\vec{v}$$

$$= \left(\hat{i}\frac{\partial}{\partial x} + \hat{j}\,\frac{\partial}{\partial y} + \,\hat{k}\frac{\partial}{\partial z}\,\right)\cdot u$$

$$+ \left(\hat{i}\frac{\partial}{\partial x} + \,\hat{j}\frac{\partial}{\partial y} + \,\hat{k}\frac{\partial}{\partial z}\right)\cdot \vec{v}$$

$$= \nabla \cdot \vec{u} + \nabla \cdot \vec{v}$$

(ii) $\nabla \cdot (f\vec{u}) = \hat{i}\,\frac{\partial}{\partial x}\,\cdot (f\,\vec{u}) + \,\hat{j}\frac{\partial}{\partial y}\cdot (f\,\vec{u}) + \vec{k}\,\frac{\partial}{\partial z}\cdot (f\,\vec{u})$

$$= \hat{i}\cdot\,\frac{\partial}{\partial x}\,(f\,\vec{u}) + \hat{j}\cdot\,\frac{\partial}{\partial y}\,(f\,\vec{u}) + \hat{k}\cdot\,\frac{\partial}{\partial z}\,(f\,\vec{u})$$

$$= \sum \hat{i}\,\left[\left(\frac{\partial f}{\partial x}\right)\vec{u} + f\left(\frac{\partial \vec{u}}{\partial x}\right)\right]$$

$$= \sum \hat{i}\,\frac{\partial f}{\partial x}\cdot\,\vec{u} + f\sum \hat{i}\cdot\,\frac{\partial \vec{u}}{\partial x}$$

$$= (\nabla f)\cdot \vec{u} + f\,(\nabla \cdot \vec{u})$$

2. (i) $\nabla\,\times (f\vec{u}) = \left(\hat{i}\frac{\partial}{\partial x} + \,\hat{j}\,\frac{\partial}{\partial y} + \,\hat{k}\,\frac{\partial}{\partial z}\right)\times f\vec{u}$

$$= \sum \hat{i}\,\frac{\partial}{\partial x}\,\times f\vec{u}$$

$$= \sum \hat{i}\,\times\,\frac{\partial}{\partial x}\,(f\vec{u})$$

$$= \sum \hat{i}\,\times\left(\frac{\partial f}{\partial x}\,\vec{u} + f\,\frac{\partial \vec{u}}{\partial x}\,\right)$$

$$= \sum \hat{i}\,\times\,\frac{\partial f}{\partial x}\,\vec{u} + \sum \hat{i}\,\times f\frac{\partial \vec{u}}{\partial x}$$

$$= \left(\sum \hat{i}\,\frac{\partial f}{\partial x}\right)\times \vec{u} + f\left(\sum \hat{i}\,\frac{\partial \vec{u}}{\partial x}\,\right)$$

$$= (\nabla f)\times \vec{u} + f\,(\nabla \times \vec{u})$$

(ii) $\nabla \times (\overrightarrow{u} \times \overrightarrow{v}) = \sum \hat{i} \frac{\partial}{\partial x} \times (\overrightarrow{u} \times \overrightarrow{v})$

$$= \sum \hat{i} \times \frac{\partial}{\partial x} (\overrightarrow{u} \times \overrightarrow{v})$$

$$= \sum \hat{i} \times \left(\frac{\partial \overrightarrow{u}}{\partial x} \times \overrightarrow{v} + \overrightarrow{u} \times \frac{\partial \overrightarrow{v}}{\partial x} \right)$$

$$= \sum \hat{i} \times \left(\frac{\partial \overrightarrow{u}}{\partial x} \times \overrightarrow{v} \right) + \sum \hat{i} \times \left(\overrightarrow{u} \times \frac{\partial \overrightarrow{v}}{\partial x} \right)$$

Now

$$\sum \hat{i} \times \left(\frac{\partial \overrightarrow{u}}{\partial x} \times \overrightarrow{v} \right) = \sum \left\{ \left(\hat{i} \cdot \overrightarrow{v} \right) \frac{\partial \overrightarrow{u}}{\partial x} - \left(\hat{i} \cdot \frac{\partial \overrightarrow{u}}{\partial x} \right) \overrightarrow{v} \right\}$$

$$\because \overrightarrow{a} \times \left(\overrightarrow{b} \times \overrightarrow{c} \right) = \overrightarrow{b} \left(\overrightarrow{a} \cdot \overrightarrow{c} \right) - \left(\overrightarrow{a} \cdot \overrightarrow{b} \right) \overrightarrow{c}$$

$$\sum \left(\hat{i} \cdot \overrightarrow{v} \right) \frac{\partial \overrightarrow{u}}{\partial x} = \left(\hat{i} \cdot \overrightarrow{v} \right) \frac{\partial \overrightarrow{u}}{\partial x} + \left(\hat{j} \cdot \overrightarrow{v} \right) \frac{\partial \overrightarrow{u}}{\partial y} + \left(\hat{k} \cdot \overrightarrow{v} \right) \frac{\partial \overrightarrow{u}}{\partial z}$$

$$= v_1 \frac{\partial u}{\partial x} + v_2 \frac{\partial u}{\partial y} + v_3 \frac{\partial u}{\partial z}$$

where $\overrightarrow{v} = v_1 \hat{i} + v_2 \hat{j} + v_3 \hat{k}$

$$= \left(v_1 \frac{\partial}{\partial x} + v_2 \frac{\partial}{\partial y} + v_3 \frac{\partial}{\partial u} \right) \overrightarrow{u}$$

$$\sum \left(\hat{i} \cdot \overrightarrow{v} \right) \frac{\partial \overrightarrow{u}}{\partial x} = \left\{ \left(v_1 \hat{i} + v_2 \hat{j} + v_3 \hat{k} \right) \cdot \left(\hat{i} \frac{\partial}{\partial x} + \hat{j} \frac{\partial}{\partial y} + \hat{k} \frac{\partial}{\partial z} \right) \right\} \overrightarrow{u}$$

$$= (\overrightarrow{v} \cdot \nabla) \overrightarrow{u}$$

$$\sum \left(\hat{i} \cdot \frac{\partial \overrightarrow{u}}{\partial x} \right) \overrightarrow{v} = \left(\sum \hat{i} \cdot \frac{\partial \overrightarrow{u}}{\partial x} \right) \overrightarrow{v}$$

$$= \left(\sum \hat{i} \frac{\partial}{\partial x} \cdot \overrightarrow{u} \right) \overrightarrow{v}$$

$$= \left\{ \left(\hat{i} \frac{\partial}{\partial x} + \hat{j} \frac{\partial}{\partial y} + \hat{k} \frac{\partial}{\partial z} \right) \cdot \overrightarrow{u} \right\} \overrightarrow{v}$$

$$= (\nabla \cdot \overrightarrow{u}) \overrightarrow{v}$$

$$\therefore \sum \hat{i} \times \left(\frac{\partial \overrightarrow{u}}{\partial x} \times \overrightarrow{v} \right) = (\overrightarrow{v} \cdot \nabla) \overrightarrow{u} - (\nabla \cdot \overrightarrow{u}) \overrightarrow{v}$$

Also

$$\sum \hat{i} \times \left(\vec{u} \times \frac{\partial \vec{v}}{\partial x}\right) = -\sum \hat{i} \times \left(\frac{\partial \vec{v}}{\partial x} \times \vec{u}\right)$$
$$= -(\vec{u} \cdot \nabla)\, \vec{v} + (\nabla \cdot \vec{v})\, \vec{u}$$

(Similar to the previous result)
Hence

$$\nabla \times (\vec{u} \times \vec{v}) = (\vec{v} \cdot \nabla)\, \vec{u} - (\nabla \cdot \vec{u})\, \vec{v} - (\vec{u} \cdot \nabla)\, \vec{v} + (\nabla \cdot \vec{v})\, \vec{u}$$
$$= (\nabla \cdot \vec{v})\, \vec{u} - (\nabla \cdot \vec{u})\, \vec{v} + (\vec{v} \cdot \nabla)\, \vec{u} - (\vec{u} \cdot \nabla)\, \vec{v}$$

Illustration 3.38: Prove that $\vec{V} = r^n\, \vec{r}$ is irrotational. Find n when it is also solenoidal.

Solution:

$$\nabla \times \vec{V} = \nabla \times (r^n\, \vec{r})$$
$$= \nabla (r^n) \times \vec{r} + r^n\, (\nabla \times \vec{r})$$

using $\nabla \times (\phi\, \vec{u}) = \left(\vec{V}\, \phi\right) \times \vec{u} + \phi\,(\nabla \times \vec{u})$

$$\nabla (r^n) = \sum \hat{i}\, \frac{\partial}{\partial x}\, r^n = \sum \hat{i}\, n r^{n-1}\, \frac{\partial \vec{r}}{\partial x}$$
$$= \sum \hat{i}\, n r^{n-1} \left(\frac{x}{r}\right) \quad \text{since } r^2 = x^2 + y^2 + z^2$$
$$= n \sum \hat{i}\, r^{n-2}\, x$$
$$\therefore\ \nabla \left(r^2\right) = n r^{n-2} \sum \hat{i}\, x = n r^{n-2} \left(x\hat{i} + y\hat{i} + z\hat{k}\right)$$
$$= n r^{n-2}\, \vec{r}$$

Since $\nabla \times \vec{r} = 0$
Hence $\nabla \times \vec{V} = n r^{n-2}\, \vec{r} \times \vec{r} + r^n\ (0)$

$$= 0, \text{ since } \vec{r} \times \vec{r} = 0$$

Hence the vector is irrotational

$$\nabla \cdot \vec{V} = \nabla \cdot (r^n\, \vec{r})$$
$$= r^n\ (\nabla \cdot \vec{r}) + \nabla (r^n) \cdot \vec{r}$$

using $\nabla \cdot (\phi \vec{u}) = \phi (\nabla \cdot \vec{u}) + \nabla \phi \cdot \vec{u}$
since $\nabla \cdot \vec{r} = 3$ and $\nabla r^n = n r^{n-2} \vec{r}$
Hence

$$\begin{aligned} \nabla \cdot \vec{V} &= 3r^n + nr^{n-2} \vec{r} \cdot \vec{r} \\ &= 3r^n + nr^{n-2} r^2 \\ &= (n+3) r^n \end{aligned}$$

If the vector $\vec{V}$ is solenoidal $\nabla \cdot \vec{V} = 0$
$\therefore (n+3)\ r^n = 0$, i.e., $n = -3$.

3.7.3.2 Formulae for grad, div, curl Involving Operator ∇ Twice

Given a scalar function f and a vector function $\vec{V}$, the following combinations of ∇ twice are possible:

(i) div (grad f) $= \nabla \cdot \nabla f = \frac{\partial^2 f}{\partial x^2} + \frac{\partial^2 f}{\partial y^2} + \frac{\partial^2 f}{\partial z^2}$

(ii) $(\nabla \cdot \nabla)\ \vec{V} = \nabla^2 \vec{V}$

(iii) grad (div $\vec{V}$) $= \nabla \left(\nabla \cdot \vec{V}\right)$

(iv) curl (grad f) $= \nabla \times \nabla f = 0$

(v) div curl $\vec{V} = \nabla \cdot \left(\nabla \times \vec{V}\right) = 0$

(vi) curl curl $\vec{V} = \nabla \times \left(\nabla \times \vec{V}\right) = \nabla \left(\nabla \cdot \vec{V}\right) - \nabla^2 \vec{V}$.

Proof:

(i) $\nabla \cdot \nabla \phi = \left(\hat{i} \frac{\partial}{\partial x} + \hat{j} \frac{\partial}{\partial y} + \hat{k} \frac{\partial}{\partial z}\right) \cdot \left(\hat{i} \frac{\partial f}{\partial x} + \hat{j} \frac{\partial f}{\partial y} + \hat{k} \frac{\partial f}{\partial z}\right)$

$$\therefore \nabla \cdot \nabla \phi = \nabla^2 \phi = \frac{\partial^2 f}{\partial x^2} + \frac{\partial^2 f}{\partial y^2} + \frac{\partial^2 f}{\partial z^2}$$

The operator $\nabla^2 = \frac{\partial^2}{\partial x^2} + \frac{\partial^2}{\partial y^2} + \frac{\partial^2}{\partial z^2}$ is called Laplace's operator or Laplacian and $\nabla^2 f = 0$ is called Laplace's equation. which frequently occurs in physical and engineering problems.

(ii) $(\nabla \cdot \nabla)\ \vec{V} = \nabla^2\vec{V} = \frac{\partial^2 \vec{V}}{\partial x^2} + \frac{\partial^2 \vec{V}}{\partial y^2} + \frac{\partial^2 \vec{V}}{\partial z^2}$

(iii) $\nabla\ (\nabla \cdot\ \nabla) = \nabla\left(\frac{\partial V_1}{\partial x} + \frac{\partial V_2}{\partial y} + \frac{\partial V_3}{\partial z}\right) =$ which is a vector

(iv) curl grad $f = \nabla\ \times \nabla f = \begin{vmatrix} \hat{i} & \hat{j} & \hat{k} \\ \frac{\partial}{\partial x} & \frac{\partial}{\partial y} & \frac{\partial}{\partial z} \\ \frac{\partial f}{\partial x} & \frac{\partial f}{\partial y} & \frac{\partial f}{\partial z} \end{vmatrix} = 0$ on simplification

(v) div curl $\vec{V} = \nabla \cdot (\nabla\ \times V)$

$$\nabla\ \times \vec{V} = \begin{vmatrix} \hat{i} & \hat{j} & \hat{k} \\ \frac{\partial}{\partial x} & \frac{\partial}{\partial y} & \frac{\partial}{\partial z} \\ V_1 & V_2 & V_3 \end{vmatrix}, \text{ where } \vec{V} = V_1\hat{i} + V_2\hat{j} + V_3\hat{k}$$

$$\therefore \nabla \times \vec{V} = \hat{i}\left(\frac{\partial V_3}{\partial y} - \frac{\partial V_2}{\partial z}\right) + \hat{j}\left(\frac{\partial V_1}{\partial z} - \frac{\partial V_3}{\partial x}\right) + \hat{k}\left(\frac{\partial V_2}{\partial x} - \frac{\partial V_1}{\partial y}\right)$$

$\therefore$ div curl $\vec{V} = \nabla \cdot \left(\nabla\ \times \vec{V}\right)$

$$= \frac{\partial}{\partial x}\left(\frac{\partial V_3}{\partial y} - \frac{\partial V_2}{\partial z}\right) + \frac{\partial}{\partial y}\left(\frac{\partial V_1}{\partial z} - \frac{\partial V_3}{\partial x}\right) + \frac{\partial}{\partial z}\left(\frac{\partial V_2}{\partial x} - \frac{\partial V_1}{\partial y}\right)$$

$$= \left(\frac{\partial^2 V_3}{\partial x \partial y} - \frac{\partial^2 V_2}{\partial x \partial y}\right) + \left(\frac{\partial^2 V_1}{\partial y \partial z} - \frac{\partial^2 V_3}{\partial y \partial x}\right) + \left(\frac{\partial^2 V_2}{\partial z \partial x} - \frac{\partial^2 V_1}{\partial z \partial y}\right)$$

$$= 0$$

(vi) We have $\vec{a} \times \left(\vec{b} \times \vec{c}\right) = \vec{b}\left(\vec{a} \cdot \vec{c}\right) - \left(\vec{a} \cdot \vec{b}\right)\vec{c}$

Treating ∇ as a vector if we take

$\vec{a} = \nabla$, $\vec{b} = \nabla$, and $\vec{c} = \vec{V}$, we have

$$\nabla \times \left(\nabla \times \vec{V}\right) = \nabla\ (\nabla \cdot \nabla) - \ (\nabla \cdot \nabla)\ V$$

$$\therefore \quad \nabla\ \times (\nabla\ \times V) = \nabla\ \left(\nabla \cdot \vec{V}\right) - \nabla^2\ \vec{V}$$

i.e., curl curl $\vec{V}$ = grad div $\vec{V} - \nabla^2\vec{V}$

Illustration 3.39: If $\rho \vec{E} = \nabla\phi$, where $\rho,\ \phi$ are scalar fields and $\vec{E}$ is a vector field, prove that $\vec{E}\cdot$ curl $\vec{E} = 0$.

Solution:

$$\vec{E} = \frac{1}{\rho}\,\nabla\phi$$

$$\text{curl}\vec{E} = \nabla\times\vec{E} = \nabla\times\left(\frac{1}{\rho}\nabla\phi\right)$$

$$= \nabla\left(\frac{1}{\rho}\right)\times\nabla\phi + \frac{1}{\rho}\,\nabla\ \times(\nabla\phi)$$

$$[\because\ \nabla\times(f\vec{u}) = \nabla f\times\vec{u} + f\,(\nabla\times\vec{u})$$

$$= \nabla\left(\frac{1}{\rho}\right)\times\ \nabla\phi + 0\,[\because\ \nabla\ \times\ \nabla f = 0$$

$$\therefore\ \vec{E}\ \text{curl}\vec{E} = \vec{E}\times\left[\nabla\left(\tfrac{1}{\rho}\right)\times\nabla\phi\right]$$

$$= \nabla\left(\frac{1}{\rho}\right)\cdot\left[\nabla\phi\ \times\vec{E}\right]$$

$$[\because\vec{a}\cdot\left(\vec{b}\times\vec{c}\right) = \vec{b}\cdot(\vec{c}\times\vec{a})$$

$$= \nabla\left(\frac{1}{\rho}\right)\cdot\left[\rho\,\vec{E}\times\vec{E}\right]$$

$$= 0\ \ \text{since}\,\vec{E}\times\vec{E} = 0$$

Illustration 3.40: Prove that $\nabla^2 f\,(r) = f''(r) + \frac{2}{r}\,f'(r)$.

Solution:

$$\nabla^2 f\,(r) = \nabla\cdot\{\nabla\ f\,(r)\} = \text{div}\,\{\text{grad}\ f\,(r)\}$$

$$= \left(\frac{\partial^2}{\partial x^2} + \frac{\partial^2}{\partial y^2} + \frac{\partial^2}{\partial z^2}\right) f\,(r)$$

Now $\frac{\partial}{\partial x}\ f\,(r) = \frac{\partial}{\partial x}\ f\,(r)\cdot\frac{\partial r}{\partial x} = f'\,(r)\cdot\frac{x}{r}$ since $r^2 = x^2 + y^2 + z^2$

$$\therefore\ \frac{\partial^2}{\partial x^2}\ f\,(r) = \frac{\partial}{\partial x}\ \left[f'\,(r)\cdot x\cdot r^{-1}\right]$$

$$= f'\,(r)\,r^{-1}\cdot 1 + f''(r)\ \cdot\ \frac{x}{r}\cdot\ x\,r^{-1} - f'(r)\,x^2 r^{-3}$$

$$= \frac{f'\,(r)}{r} + \frac{x^2 f''(r)}{r^2} - \frac{x^2 f'\,(r)}{r^3}$$

Similarly,

$$\frac{\partial^2}{\partial y^2} f(r) = \frac{f'(r)}{r} + \frac{y^2 f''(r)}{r^2} - \frac{y^2 f'(r)}{r^3}$$

and

$$\frac{\partial^2}{\partial z^2} f(r) = \frac{f'(r)}{r} + \frac{z^2 f''(r)}{r^2} - \frac{z^2 f'(r)}{r^3}$$

Adding, we have

$$\left(\frac{\partial^2}{\partial x^2} + \frac{\partial^2}{\partial y^2} + \frac{\partial^2}{\partial z^2}\right) f(r) = \frac{3f'(r)}{r} + f''(r) - \frac{f'(r)}{r}$$

$$= \frac{2f'(r)}{r} + f''(r)$$

$\therefore\ \nabla^2 f(r) = \text{div grad } f(r) = \frac{2f'(r)}{r} + f''(r)$.

Illustration 3.41: Maxwell's equations of the electromagnetic theory are

$$\nabla \cdot \vec{E} = 0,\ \nabla \cdot \vec{H} = 0,\ \nabla \times \vec{E} = -\frac{\partial \vec{H}}{\partial t},\ \ \nabla \times \vec{H} = \frac{\partial \vec{E}}{\partial t}$$

Show that $\vec{E}$ and $\vec{H}$ satisfy wave equations

(i) $\nabla^2 \vec{E} = \frac{\partial^2 \vec{E}}{\partial t^2}$ and (ii) $\nabla^2 \vec{H} = \frac{\partial^2 \vec{H}}{\partial t^2}$

Solution:

$$\nabla \times \left(\nabla \times \vec{E}\right) = \nabla \times \left(-\frac{\partial \vec{H}}{\partial t}\right) = -\frac{\partial}{\partial t}\left(\nabla \times \vec{H}\right)$$

$$= -\frac{\partial}{\partial t}\left(\frac{\partial E}{\partial t}\right) = -\frac{\partial^2 E}{\partial t^2} \qquad \text{(i)}$$

But

$$\nabla \times \left(\nabla \times \vec{E}\right) = \nabla\left(\nabla \cdot \vec{E}\right) - \nabla^2 \vec{E} = -\nabla^2 \vec{E} \qquad \text{(ii)}$$

Then from (i) are (ii), we have

$$\nabla^2 \vec{E} = \frac{\partial^2 \vec{E}}{\partial t^2}$$

Now

$$\nabla \times \left(\nabla \times \vec{H}\right) = \nabla \times \left(\frac{\partial \vec{E}}{\partial t}\right) = \frac{\partial}{\partial t}\left(\nabla \times \vec{E}\right)$$

$$= \frac{\partial}{\partial t}\left(-\frac{\partial \vec{H}}{\partial t}\right) = -\frac{\partial^2 \vec{E}}{\partial t^2} \qquad \text{(iii)}$$

but

$$\nabla \times \left(\nabla \times \vec{H}\right) = \nabla\left(\nabla \cdot \vec{H}\right) - \nabla^2\vec{H} = -\nabla^2\vec{H} \qquad \text{(iv)}$$

from (iii) and (iv), we have

$$\nabla^2\vec{H} = \frac{\partial^2 \vec{H}}{\partial t^2}$$

The equation $\nabla^2 \mathrm{u} = \frac{\partial^2 u}{\partial t^2}$
i.e., $\frac{\partial^2 u}{\partial x^2} + \frac{\partial^2 u}{\partial y^2} + \frac{\partial^2 u}{\partial z^2} = \frac{\partial^2 u}{\partial t^2}$ is called the wave equation.

3.8 Exercise

1. Determine k and τ for the following curve:

   (i) $x = t\cos t\ ,\ y = t\sin t\ ,\ z = \lambda t$ at $t = 0$

   $\left(\text{Answer: } k = \frac{2}{1+\lambda^2},\ \tau = \frac{3\lambda}{2(1+\lambda^2)}\right)$

   (ii) $x = 3t,\ y = 3t^2,\ z = 2t^3$

   $\left(\text{Answer:} \varrho = \sigma = \frac{3}{2}\left(1 + 2t^2\right)^2, \varrho = \frac{1}{k},\ \sigma = \frac{1}{\tau}\right)$

   (iii) $x = a\left(3t - t^3\right),\ y = 3\,at^2,\ z = a(3t + t^3)$

   $\left(\text{Answer: } \varrho = \sigma = 3a\left(1 + t^2\right)^2\right)$

2. For the curve $x = 3t,\ y = 3t^2,\ z = 2t^2$ at $t = 1$ Prove that $\varrho = \sigma = \frac{3}{2}\left(1 + 2t^2\right)^2$.

3. For the curve $x = 2$ log t, $y = 4t,\ z = 2t^2 + 1$ at $t = 1$ prove that $\varrho = \sigma = 9$.

4. Find curvature and torsion for the curve $\vec{r} = \cos t\ \hat{i} + \sin t\,\hat{j} + t\hat{k}$. Also, prove that $2\left(\kappa^2 + \tau^2\right) = 1$.

   $\left(\text{Answer: } k = \frac{1}{2},\ \tau = \frac{1}{2}\right)$

5. Find curvature and torsion for the curve $x = t\cos t$, $y = t\sin t$, $z = \lambda t$ at $t = 0$.

$$\left(\text{Answer: } k = \frac{2}{1+\lambda^2},\ \tau = \frac{3\lambda}{2(1+\lambda^2)}\right)$$

6. For the curve $x = a\cos\theta$, $y = a\sin\theta$, $z = a\,\theta\tan\alpha$, find ρ.

$$(\text{Answer:} \rho = a\sec^2\alpha\ )$$

7. For the curve $x = a\cos\theta$, $y = a\sin\theta$, $z = \lambda\theta$, find k and τ.

$$\left(\text{Answer: } k = \frac{a}{a^2+\lambda^2},\ \tau = \frac{\lambda}{a^2+\lambda^2}\right)$$

8. Show that for the curve

$$x^2 - y^2 = c^2,\ y = x\tanh\frac{z}{c}\ ,\ \varrho = \sigma = \frac{2x^2}{c}.$$

9. For the curve $x = \tan^{-1}s$, $y = \frac{1}{\sqrt{2}}\log\left(s^2+1\right)$, $z = s - \tan^{-1}s$

show that $k = \tau = \frac{\sqrt{2}}{s^2+1}$.

10. For the curve $x = a\left(3u - u^3\right)$, $y = 3au^2$, $z = a\left(3u + u^3\right)$

show that $k = \tau = \frac{1}{3a(1+u^2)^2}$.

11. For the curve $x = 4a\cos^3 u$, $y = 4a\sin^3 u$, $z = 3c\cos 2u$

prove that $k = \frac{a}{6(a^2+c^2)\sin 2u}$.

12. Find the length of the curve $x = e^t\cos t$, $y = e^t\sin t$, $z = e^t$ between $t = t_1$ and $t = t_2$.

$$\left(\text{Answer: } \sqrt{3}\left(e^{t_2} - e^{t_1}\right)\right)$$

13. Find the length of the arc of the curve $x = 3t$, $y = 3t^2$, $z = 2t^3$ between $t = 0$ and $t = 1$.

(Answer: 5)

14. Find the length of the curve $x = a\cos^3 t$, $y = a\sin^3 t$, $z = \frac{3a}{2}\cos^2 t$ from the point $\left(a,\ 0,\ \frac{3a}{2}\right)$ to the point $(0,\ a,\ 0)$.

$$\left(\text{Answer: } \frac{3a}{\sqrt{2}}\right)$$

15. Find the length of the curves $\vec{r}\,(t) = 2t\hat{i} + 3\sin 2t\,\hat{j} + 3\cos 2t\,\hat{k}$ on the interval $0 \le t \le 2\pi$.

(Answer: $2\sqrt{10t}$)

16. Find the magnitude of the velocity and acceleration of a particle which moves along the curve $x = 2\sin 3t$, $y = 2\cos 3t$, $z = 8t$ at any time $t > 0$. Find the unit tangent vector to the curve.

$\left(\text{Answer: } 10, 18, \frac{1}{10}\left[(6\cos 3t)\,\hat{i} - (6\sin 3t)\,\hat{j} + 8\hat{k}\right]\right)$

17. A particle moves along a plane curve such that its linear velocity is perpendicular to the radius vector. Show that the path of the particle is a circle.

18. Find the magnitude of the tangential components of acceleration at any time t of a particle whose position at any time t is given by $x = \cos t + t\sin t$, $y = \sin t - t\cos t$.

(Answer: 1)

19. Show that the length of the curve $2x = a\,(\cos 3\theta + \cos\theta)$, $2y = a\,(\sin 3\theta + \sin\theta)$ $z = \sqrt{3}\ a\cos\theta$ measured from the point $\left(a,\ 0,\ \sqrt{3}\,a\right)$ in $2a\theta$.

20. Prove that $\frac{d}{dt}\left(V.\,\frac{dV}{dt} \times \frac{d^2V}{dt^2}\right) = V\,.\,\frac{dV}{dt} \times \frac{d^3V}{dt^3}$.

21. Show that $r = e^{-1}\,(a\cos 2t + b\sin 2t)$, where a and b are constant vectors, is a solution to the differential equation $\frac{d^2r}{dt^2} + 2\frac{dr}{dt} + 5r = 0$.

22. What is the greatest rate of increase of $u = x^2 + yz^2$ at the point $(1, -1, 3)$?

$\left(\text{Answer: } \sqrt{89}\right)$

23. Find grad ϕ at the point $(1, -2, 1)$ when ϕ is given by $\phi = 3x^2y - y^3z^2$.

$\left(\text{Answer: } -12\hat{i} - 9\hat{j} - 16\hat{k}\right)$

24. If $\phi = x^3 + y^3 + z^3 - 3xyz$, show that $\vec{r} \cdot \nabla\phi = 3\phi$.

25. If $u = x + y + z,\ v = x^2 + y^2 + z^2,\ \ w = yz + zx + xy$ then

   a. Prove that $(\nabla\, u) \cdot [\nabla v \times \nabla w] = 0$.

b. Show that $\nabla u, \ \nabla v, \ \nabla w$ are coplanar.

26. If $\nabla\phi = (2xyz)\,\hat{\imath} + \left(x^2 z\right)\hat{\jmath} + \left(x^2 y\right)\hat{k}$, determine ϕ.

(Answer: $\phi = x^2 yz + c$)

27. If $\overrightarrow{A}$ is a constant vector then prove that $\nabla\left(r \cdot \overrightarrow{A}\right) = \overrightarrow{A}$.

28. Find a unit vector normal to the surface $x^2 + y^2 + z^2 = a^2$ at $\left(\frac{a}{\sqrt{3}}, \frac{a}{\sqrt{3}}, \frac{a}{\sqrt{3}}\right)$.

$\left(\text{Answer: } \frac{\hat{\imath}+\hat{\jmath}+\hat{k}}{\sqrt{3}}\right)$

29. Find a unit vector normal to the surface $x^2 + y^2 - z = 1$ at the point $(1, 1, 1)$.

$\left(\text{Answer: } \frac{2\hat{\imath}-2\hat{\jmath}-\hat{k}}{3}\right)$

30. Find a unit outward drawn normal to the surface $\left(x^2 - 1\right) + y^2 + (z+2)^2 = 9$ at the point $(3, 1, -4)$.

$\left(\text{Answer: } \frac{2\hat{\imath}+\hat{\jmath}-2\hat{k}}{3}\right)$

31. The temperature at a point (x, y, z) in space is given by $T(x, y, z) = x^2 + y^2 - z$. A mosquito located at $(1, 1, 2)$ desires to fly in such a direction that it will get warm as soon as possible. In what direction should it fly?

$\left(\text{Answer: } \frac{1}{3}\left(2\hat{\imath} + 2\hat{\jmath} - \hat{k}\right)\right)$

32. What is the angle between the normals to the surface $xy = z^2$ at the points $(1, 9, -3)$ and $(-2, -2, 2)$?

$\left(\text{Answer: } \cos^{-1}\left(\frac{11}{\sqrt{118}}\right)\right)$

33. Show that $\nabla\phi$ is a vector perpendicular to the surface $\phi(x, y, z) = 0$.

34. Find the directional derivative of $\phi(x, y, z) = xy^2 + yz^3$ at the point $(2, -1, 1)$ in the direction of the vector $\hat{\imath} + 2\hat{\jmath} + 2\hat{k}$.

(Answer: $-\frac{11}{3}$)

35. Find the directional derivative of $\phi(x, y, z) = xy^2 + yz^2$ at the point $(2, -1, 1)$ in the direction of the vector $\hat{i} + 2\hat{j} + 2\hat{k}$.

(Answer: -3)

36. If the directional derivative of $\phi = axy + byz + czx$ at $(1,\ 1,\ 1)$

has maximum magnitude 4 in a direction parallel to the x-axis, find the values of $a,\ b,\ c$.

(Answer: $a = 2,\ b = -2,\ c = 2$)

37. Prove that the angle between the surface $x^2 + y^2 + z^2 = 9$ and

$x^2 + y^2 - z = 3$ at the point $(2,\ -1,\ 2)$ is $\cos^{-1}\left(\frac{8}{3\sqrt{21}}\right)$.

38. Find the angle between the tangent planes to the surfaces $x\log z = y^2 - 1$ and $x^2 y = 2 - z$ at the point $(1,\ 1,\ 1)$.

$\left(\text{Answer: } \cos^{-1}\left(\frac{1}{\sqrt{30}}\right)\right)$

39. In what direction form $(3,\ 1,\ -2)$ is the directional derivative of $\phi = x^2\ y^2\ z^4$ maximum, and what is its magnitude?

$\left(\text{Answer: } \frac{1}{13}\left(4\hat{i} + 3\hat{j} - 12\hat{k}\right), 1\right)$

40. Is there a direction u in which the rate of change of $f(x,\ y) = x^2 - 3xy + 4y^2$ at $P(1,\ 2)$ equals 14? Give reasons for your answer.

$\left(\text{Answer : No, the maximum rate of change is } \sqrt{185} < 14\right)$

41. Given $\vec{u} = xyz\ \hat{i} + \left(2x^2 z - y^2 x\right)\ \hat{j} + xz^3\ \hat{k},\ \vec{v} = x^2\hat{i} + 2yz\hat{j} + (1 + 2z)\ \hat{k}, f = xy + yz + z^2$ then find (i) $\nabla \cdot \vec{v}$ (ii) $\nabla \times v$ (iii) $\nabla \cdot (f\vec{u})$ (iv) $\nabla \times (f\vec{v})$ at $(1,\ 0,\ -1)$

$\left(\text{Answer: (i) } 3,\ \text{(ii) } 0,\ \text{(iii) } 5,\ \text{(iv) } -2\hat{j}\right)$

42. If ω is a constant vector and $\vec{v} = \vec{\omega} \times \vec{r}$ prove that div $\vec{v} = 0$.

43. Prove that $\vec{F} = \frac{x\hat{i}+y\hat{j}}{x^2+y^2}$ is solenoidal.

44. If $\vec{F} = \left(y^2 - z^2 + 3yz - 2x\right)\hat{i} + (3xz + 2xy)\hat{j} + (3xy - 2xz + 2z)\ \hat{k}$, Show that $\vec{F}$ is both solenoidal and irrotational.

45. If $\vec{F} = (z^2 + 2x + 3y)\ \hat{\imath} + (3x + 2y + z)\ \hat{\jmath} + (y + 2xz)\ \hat{k}$, Show that F is irrotational but not solenoidal.

46. Show that $\vec{F} = (6xy + z^3)\ \hat{\imath} + (3x^2 - z)\ \hat{\jmath} + (3xz^2 - y)\ \hat{k}$

is irrotational. Find scalar ϕ such that $\vec{F} = \nabla\phi$.

47. In each case the velocity vector $\vec{v}$ of a steady fluid motion is given, Find curl $\vec{v}$. Is the motion incompressible?

(i) $\vec{v} = z^2\hat{\jmath}$, (ii) $\vec{v} = y\hat{\imath} - x\hat{\jmath}$, (iii) $\vec{v} = x\hat{\imath} + y\hat{\jmath}$

$$\left(\begin{array}{ll}\text{Answer:} & \text{(i) curl } \vec{v} = -2z\hat{\jmath}, \text{ incompressible} \\ & \text{(ii) curl } \vec{v} = -2\hat{k}, \text{ incompressible} \\ & \text{(iii) curl } \vec{v} = 0, \text{ div } \vec{v} = 2, \text{compressible}\end{array}\right)$$

48. If the vector product of the vectors $\vec{A}$ and $\vec{B}$ be the curl of a third vector, prove that $\vec{A}\cdot$ curl $= \vec{B}\cdot$ curl $\vec{A}$.

49. Prove that div (f curl $\vec{F}$) = (grad f) · curl $\vec{F}$.

50. If $\phi,\ \psi$ satisfy Laplace equation, prove that the vector $(\phi\nabla\psi - \psi\nabla\phi)$ is solenoidal.

51. Show that

(i.) $\nabla^2\left[\nabla.\left(\frac{\mathrm{r}}{\mathrm{r}^2}\right)\right] = \frac{2}{\mathrm{r}^4}$

(ii.) $\nabla \times \left(a \times \nabla\frac{1}{r}\right) + \nabla\left(a\ .\ \nabla\frac{1}{r}\right) = 0.$

52. If $\frac{d\vec{u}}{dt} = \vec{w} \times \vec{u}$ and $\frac{d\vec{v}}{dt} = \vec{w} \times \vec{v}$, then prove that

$$\frac{d}{dt}\left(\vec{u} \times \vec{v}\right) = \vec{w} \times \left(\vec{u} \times \vec{v}\right).$$

53. If $\vec{r} = t^3\ \hat{\imath} + \left(2t^3 - \frac{1}{5t^2}\right)\hat{\jmath}$, then show that $\hat{r} \times \frac{d\vec{r}}{dt} = \hat{k}$.

4

Vector Integral Calculus

4.1 Introduction

Vector calculus deals with the differentiation and integration of vector functions. We have learned about the derivative of a vector function, gradient, divergence, and curl in vector differential calculus. In vector integral calculus, we learn about line integral, surface integral, and volume integral. It plays an important role in differential geometry and the study of partial differential equations. It is useful in the study of rigid dynamics, fluid dynamics, heat transfer, electromagnetism, theory of relativity, etc.

4.2 Line Integrals

The line integral is a simple generalization of a definite integral $\int_a^b f(x)dx$ which is integrated from $x = a$ (point A) to $x = b$ (point B) along the x-axis. In a line integral, the integration is done along a curve C in space.

Let $\overrightarrow{F}(\overrightarrow{r})$ be a vector function defined at every point of a curve C in Figure 4.1. If $\overrightarrow{r}$ is the position vector of a point $P(x, y, z)$ on the curve C, then the line integral of $\overrightarrow{F}(\overrightarrow{r})$ over a curve C is defined by

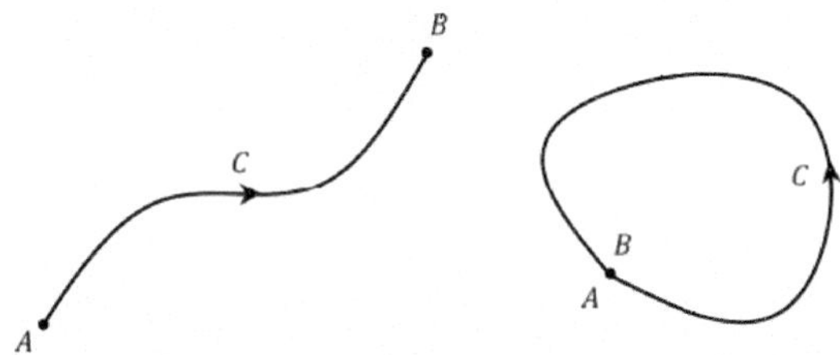

Figure 4.1 Representation of a vector function defined at every point of a curve C

$$\int_C \vec{F}(\vec{r}) \cdot d\vec{r} = \int_C F_1 dx + F_2 dy + F_3 dz$$

where, $\vec{F} = F_1\,\hat{i} + F_2\,\hat{j} + F_3\hat{k}$ and $\vec{r} = x\hat{i} + y\hat{j} + \hat{k}$

If the curve C is represented by a parametric representation

$$\vec{r}\,(t) = x\hat{i} + y\hat{j} + z\hat{k},$$

then the line integral along the curve C from $t = a$ to $t = b$ is

$$\int_C \vec{F}(\vec{r}) \cdot d\vec{r} = \int_a^b \vec{F} \cdot \frac{d\vec{r}}{dt} dt$$
$$= \int_a^b \left(F_1 \frac{dx}{dt} + F_2 \frac{dy}{dt} + F_3 \frac{dz}{dt} \right) dt$$

If C is a closed curve, then the symbol of the line integral $\int_C$ is replaced by $\oint_C$.

Two other types of line integrals are $\int_C \vec{F} \times d\vec{r}$ and $\int_C \phi d\vec{r}$, (where ϕ is a scalar point function) which are both vectors.

Note:

1. The curve C is called the path of integration, the points $\vec{r}(a)$ and $\vec{r}(b)$ are called initial and terminal points respectively.
2. The direction from A and B along which t increases is called positive direction on C.

4.2.1 Circulation

If $\vec{F}$ is the velocity of a fluid particle and C is a closed curve, then the line integral $\oint_C \vec{F} \cdot d\vec{r}$ represents the circulation of $\vec{F}$ around the curve C.

Note:

1. If the circulation of $\vec{F}$ around every closed curve C in the region R is zero, then $\vec{F}$ is irrotational, i.e., if $\oint_A^B \vec{F} \cdot d\vec{r} = 0$, $\vec{F}$ is irrotational.

4.2.2 Work Done by a Force

If $\vec{F}$ is the force acting on a particle moving along the arc AB of the curve C, then the line integral $\int_A^B \vec{F} \cdot d\vec{r}$ represents the work done in displacing the particle from the point A to the point B.

4.3 Path Independence of Line Integrals

4.3.1 Theorem: Independent of Path

The necessary and sufficient condition that the line integral $\int_A^B \overrightarrow{F} \cdot d\overrightarrow{r}$ be independent of the path is that $\overrightarrow{F}$ is the gradient of some scalar function ϕ.

Proof: Let $\overrightarrow{F} = F_1\hat{i} + F_2\hat{j} + F_3\hat{k}$ be the gradient of a scalar function ϕ (i.e., $\overrightarrow{F}$ is conservative) i.e., $\overrightarrow{F} = \nabla\phi$ in components

$$F_1 = \frac{\partial\phi}{\partial x}, \quad F_2 = \frac{\partial\phi}{\partial y}, \quad F_3 = \frac{\partial\phi}{\partial z}$$

where ϕ is a scalar potential, then the line integral along the curve C from the point A to B is

$$\begin{aligned}\int_C \overrightarrow{F} \cdot d\overrightarrow{r} &= \int_A^B \nabla\phi \cdot d\overrightarrow{r}\\ &= \int_A^B \left(\frac{\partial\phi}{\partial x}dx + \frac{\partial\phi}{\partial y}dy + \frac{\partial\phi}{\partial z}dz\right)\\ &= \int_A^B d\phi\\ &= \phi(B) - \phi(A)\end{aligned}$$

where $\phi(A)$ and $\phi(B)$ are the values of ϕ at A and B respectively, That means the line integral depends only on the start and end values of the scalar potential, not on the path of the curve. Hence the condition is necessary. We show below that the condition is sufficient.

Let $\int_A^B \overrightarrow{F} \cdot d\overrightarrow{r}$ depend on the end values A and B and not on the path of integration.

$$\therefore \int_A^B \overrightarrow{F} \cdot d\overrightarrow{r} = \phi(B) - \phi(A) = [\phi]_A^B$$

$$\therefore \overrightarrow{F} \cdot d\overrightarrow{r} = d\phi = \frac{\partial\phi}{\partial x}dx + \frac{\partial\phi}{\partial y}dy + \frac{\partial\phi}{\partial z}dz = \nabla\phi \cdot d\overrightarrow{r}$$

Since this is true for all the curves between A and B, $\overrightarrow{F} = \nabla\phi$.

In Mechanics, $\overrightarrow{F}$ is called a conservative field and ϕ is the scalar potential, and in Potential theory, if ϕ is potential and $\overrightarrow{F}$ is a force then $\overrightarrow{F} = \text{grad } \phi$. Thus, the line integral is independent of the path in C iff $\overrightarrow{F}$ is the gradient of potential in C.

Corollary 4.3.1.1: If $\overrightarrow{F} = \nabla\phi$, we have

$$\text{curl } \overrightarrow{F} = \text{curl grad } \phi = \nabla \times \nabla\phi = 0$$

Hence, the necessary and sufficient condition that $\int_C \overrightarrow{F} \cdot d\overrightarrow{r}$ be independent of the path is the curl $\overrightarrow{F}$ vanishes identically and hence $\overrightarrow{F}$ is irrotational.

Corollary 4.3.1.2: If $\int_C \overrightarrow{F} \cdot d\overrightarrow{r}$ is independent of the path of integration, then $\oint_C \overrightarrow{F} \cdot d\overrightarrow{r}$ along any closed path is zero.

Corollary 4.3.1.3: Let $\overrightarrow{F} = \hat{\imath}P + \hat{\jmath}Q$, $\int_C (Pdx + Qdy)$ is independent of its path, if $\frac{\partial P}{\partial y} = \frac{\partial Q}{\partial x}$.

Note:

1. As shown in Figure 4.2,

   if $\overrightarrow{F}$ is conservative and curve C is closed, then

$$\oint_C \overrightarrow{F} \cdot d\overrightarrow{r} = \phi(A) - \phi(A) = 0.$$

2. The work done in moving a particle from points A to B under a conservative force field is

$$\text{Work done} = \phi(B) - \phi(A)$$

.

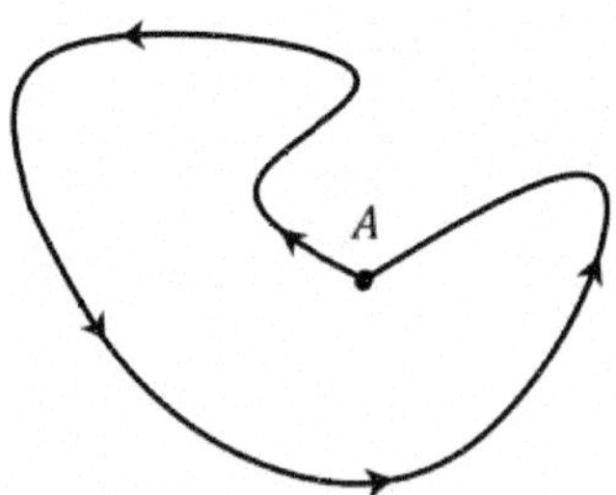

Figure 4.2 Representation of a closed curve C

Illustration 4.1: Evaluate $\int_C \overrightarrow{F} \cdot d\overrightarrow{r}$ along the parabola $y^2 = x$ between the points $(0, 0)$ and $(1, 1)$ where $\overrightarrow{F} = x^2\hat{\imath} + xy\hat{\jmath}$.

Solution:
Let $\overrightarrow{r} = x\hat{i} + y\hat{j}$

$$\therefore d\overrightarrow{r} = \hat{i}dx + \hat{j}dy$$

Here, $\overrightarrow{F} = x^2\hat{i} + xy\hat{j}$

$$\therefore \overrightarrow{F} \cdot d\overrightarrow{r} = \left(x^2\hat{i} + xy\hat{j}\right) \cdot \left(\hat{i}dx + \hat{j}dy\right)$$

$$= x^2dx + xydy \tag{4.1}$$

In the above expression, we can see that
$\overrightarrow{F} \cdot d\overrightarrow{r}$ contains both the variables x and y.

To apply line integral in Equation (4.1), we require the expression either in terms of x or in terms of y variable. As we have discussed that the line integral is independent of the path, we can consider the expression in Equation (4.1) in terms of any one variable. In this illustration, we represent the Equation (4.1) in terms of y variable and as shown in Figure 4.3, the path of integration is the parabola

$$x = y^2 \Rightarrow dx = 2ydy$$

Substituting in the Equation (4.1) and integrating between the limits $y = 0$ to $y = 1$, we get

$$\int_C \overrightarrow{F} \cdot d\overrightarrow{r} = \int_0^1 (y^4 \cdot 2ydy + y^2 \cdot ydy)$$

$$= \int_0^1 \left(2y^5 + y^3\right) dy$$

$$= \left| 2\frac{y^6}{6} + \frac{y^4}{4} \right|_0^1 = \frac{1}{3} + \frac{1}{4} = \frac{7}{12}$$

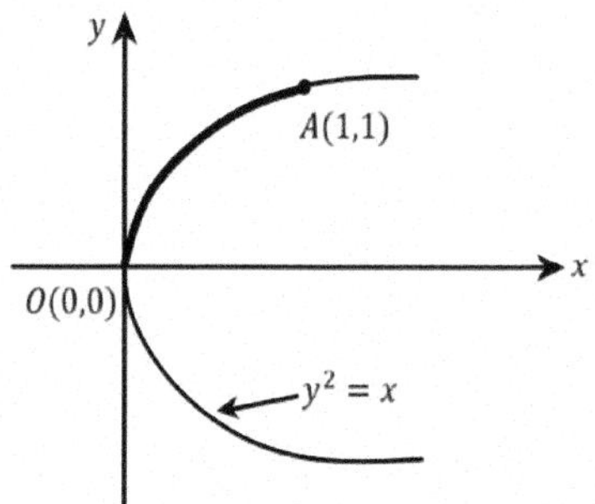

Figure 4.3 Representation of parabola $x = y^2$

Illustration 4.2: Prove that $\int_C \vec{F} \cdot d\vec{r} = 3\pi$, where $\vec{F} = z\hat{i} + x\hat{j} + y\hat{k}$ and C is the arc of the curve $\vec{r} = \text{cos}t\ \hat{i} + \text{sin}t\ \hat{j} + t\hat{k}$ from $t = 0$ to $t = 2\pi$.

Solution: Here, we have $\vec{r} = \text{cos}t\ \hat{i} + \text{sin}t\ \hat{j} + t\hat{k}$

$$\therefore x = \text{cos}t,\ y = \text{sin}t\ , z = t$$
$$\therefore dx = -\text{sin}t\ dt,\ dy = \text{cos}t\ dt,\ dz = dt$$
$$\begin{aligned}\therefore \vec{F} \cdot d\vec{r} &= \left(z\hat{i} + x\hat{j} + y\hat{k}\right) \cdot \left(\hat{i}dx + \hat{j}dy + \hat{k}dz\right) \\ &= zdx + xdy + ydz \\ &= t\left(-\text{sin}t\ \right) dt + \text{cos}t\ \cdot \text{cos}t\ dt + \text{sin}t\ dt \\ &= \left(-t\text{sin}t\ + \text{cos}^2 t\ + \text{sin}t\ \right) dt\end{aligned}$$

Now, the path of integration is the arc of the curve $\vec{r} = \text{cos}t\ \hat{i} + \text{sin}t\ \hat{j} + t\hat{k}$ from $t = 0$ to $t = 2\pi$.

$$\therefore \int_C \vec{F} \cdot d\vec{r}$$
$$= \int_0^{2\pi} \left(-t\text{sin}t\ + \text{cos}^2 t\ + \text{sin}t\ \right) dt$$
$$= -|t\left(-\text{cos}t\ \right) - \left(-\text{sin}t\ \right)|_0^{2\pi} + \int_0^{2\pi} \frac{\left(1 + \text{cos}2t\ \right)}{2} dt + |-\text{cos}t\ |_0^{2\pi}$$
$$= -\left(-2\pi\right) + \left|\frac{t}{2} + \frac{\text{sin}2t}{4}\right|_0^{2\pi} - \left(\text{cos}2\pi - \text{cos}0\ \right)$$
$$= 2\pi + \frac{2\pi}{2} = 3\pi$$

Illustration 4.3 If $\vec{F} = (2x - y + 2z)\,\hat{i} + (x + y - z)\,\hat{j} + (3x - 2y - 5z)\,\hat{k}$, calculate the circulation of $\vec{F}$ along the circle in the xy-plane of 2 units radius and center at the origin.

Solution:
The circulation is given by

$$\oint_C \vec{F} \cdot d\vec{r}$$

Let $\vec{r} = x\hat{i} + y\hat{j} + z\hat{k}$ then

$$d\vec{r} = \hat{i}dx + \hat{j}dy + \hat{k}dz$$

Now, $\vec{F} \cdot d\vec{r}$

$$= \left[(2x - y + 2z)\hat{i} + (x + y - z)\hat{j} + (3x - 2y - 5z)\hat{k}\right] \cdot (\hat{i}dx + \hat{j}dy + \hat{k}dz)$$

$$= (2x - y + 2z)\,dx + (x + y - z)\,dy + (3x - 2y - 5z)\,dz$$

Next, the path of integration is the circle in xy-plane of radius 2 units and center at the origin, i.e., $x^2 + y^2 = 4$ and in xy-plane $z = 0$. Therefore, the parametric equation of the circle is

$$x = 2\cos\theta\ , \quad y = 2\ \sin\theta$$

$$dx = -2\sin\theta\ d\theta\ , \quad dy = 2\cos\theta\ d\theta$$

For the complete circle, θ varies from 0 to 2π.
Substituting in $\vec{F} \cdot d\vec{r}$ and integrating between the limits $\theta = 0$ to $\theta = 2\pi$,

$$\therefore \text{Circulation} = \int_0^{2\pi} [(2 \cdot 2\cos\theta - 2\sin\theta)\,(-2\sin\theta d\theta)$$
$$+ (2\cos\theta + 2\sin\theta)\,(2\cos\theta\ d\theta)]$$
$$= 4\int_0^{2\pi} \left(-2\cos\theta\ \sin\theta\ + \sin^2\theta\ + \cos^2\theta\ + \cos\theta\ \sin\theta\ \right) d\theta$$
$$= 4\int_0^{2\pi} \left(1 - \frac{\sin 2\theta}{2}\right) d\theta = 4\left|\theta + \frac{\cos 2\theta}{4}\right|_0^{2\pi} = 8\pi$$

Illustration 4.4 Evaluate $\int_C \vec{F} \cdot d\vec{r}$ where $\vec{F} = \left(x^2 + y^2\right)\hat{i} - 2xy\hat{j}$ and C is the rectangle in the xy-plane bounded by $y = 0,\ x = a, y = b, x = 0$.

Solution:

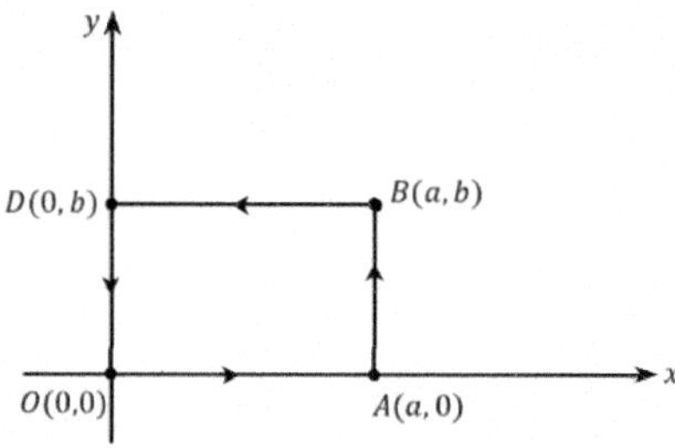

Figure 4.4 Representation of the rectangle in xy-plane bounded by lines

Consider, $\vec{r} = x\hat{i} + y\hat{j}$

$$\therefore d\vec{r} = \hat{i}dx + \hat{j}dy$$

Now, $\vec{F} \cdot d\vec{r} = \left[\left(x^2 + y^2\right)\hat{i} - 2xy\hat{j}\right] \cdot \left(\hat{i}dx + \hat{j}dy\right)$

$$= \left(x^2 + y^2\right)dx - 2xydy$$

Next, the path of integration is the rectangle $OABD$ (See Figure 4.4) bounded by the four lines $y = 0,\ x = a,\ y = b,\ x = 0$.

$$\therefore \int_C \vec{F} \cdot d\vec{r} = \int_{OA} \vec{F} \cdot d\vec{r} + \int_{AB} \vec{F} \cdot d\vec{r} + \int_{BD} \vec{F} \cdot d\vec{r} + \int_{DO} \vec{F} \cdot d\vec{r} \tag{4.2}$$

(i) Along $OA : y = 0 \Rightarrow dy = 0$ and x varies from 0 to a.

$$\int_{OA} \vec{F} \cdot d\vec{r} = \int_0^a x^2dx = \left|\frac{x^3}{3}\right|_0^a = \frac{a^3}{3}$$

(ii) Along $AB : x = a \Rightarrow dx = 0$ and y varies from 0 to b.

$$\int_{AB} \vec{F} \cdot d\vec{r} = \int_0^b -2aydy = -\left|ay^2\right|_0^b = -ab^2$$

(iii) Along $BD : y = b \Rightarrow dy = 0$ and x varies from a to 0.

$$\int_{BD} \vec{F} \cdot d\vec{r} = \int_a^0 (x^2 + y^2)dx = \left|\frac{x^3}{3} + b^2x\right|_a^0 = -\left(\frac{a^3}{3} + b^2a\right)$$

(iv) Along $DO : x = 0 \Rightarrow dx = 0$ and y varies from b to 0.

$$\int_{DO} \vec{F} \cdot d\vec{r} = \int_b^0 0dy = 0$$

Substituting in the Equation (4.2), we get

$$\int_C \vec{F} \cdot d\vec{r} = \frac{a^3}{3} - ab^2 - \frac{a^3}{3} - b^2a = -2ab^2$$

Illustration 4.5 Evaluate $\int_C \vec{F} \cdot d\vec{r}$ where $\vec{F} = \left(3x^2 + 6y\right)\hat{i} - 14yz\hat{j} + 20xz^2\hat{k}$ and C is the straight line joining the points $(0, 0, 0)$ to $(1, 1, 1)$.

Solution:
Consider, $\vec{r} = x\hat{i} + y\hat{j} + z\hat{k}$ then $d\vec{r} = \hat{i}dx + \hat{j}dy + \hat{k}dz$
Here, we have $\vec{F} = (3x^2 + 6y)\,\hat{i} - 14yz\hat{j} + 20xz^2\hat{k}$

$$\therefore \vec{F} \cdot d\vec{r} = \left[(3x^2 + 6y)\,\hat{i} - 14yz\hat{j} + 20xz^2\hat{k}\right] \cdot (\hat{i}dx + \hat{j}dy + \hat{k}dz)$$

Next, the path of integration is the straight line joining the points $A\,(0,0,0)$ to $B\,(1,1,1)$. The equation of the line AB is

$$\frac{x-0}{0-1} = \frac{y-0}{0-1} = \frac{z-0}{0-1}$$

$$\therefore x = y = z$$

$$\therefore dx = dy = dz$$

Substituting in $\vec{F} \cdot d\vec{r}$ and integrating between the limits $x = 0$ to $x = 1$, we get

$$\int_C \vec{F} \cdot d\vec{r} = \int_0^1 \left[(3x^2 + 6x)\,dx - 14x^2dx + 20x^3dx\right]$$

$$= \int_0^1 (20x^3 - 11x^2 + 6x)dx$$

$$= \left|20\frac{x^4}{4} - 11\frac{x^3}{3} + 6\frac{x^2}{2}\right|_0^1 = \frac{13}{3}$$

Illustration 4.6 If $\vec{F} = 2xyz\hat{i} + (x^2z + 2y)\,\hat{j} + x^2y\hat{k}$, then

(i) If $\vec{F}$ is conservative, then find its scalar potential ϕ.

(ii) Find the work done in moving a particle under this force field from $(0,1,1)$ to $(1,2,0)$.

Solution:

(i) Since $\vec{F}$ is conservative, we have $\vec{F} = \nabla\phi$

$$\therefore 2xyz\hat{i} + (x^2z + 2y)\,\hat{j} + x^2y\hat{k} = \hat{i}\frac{\partial\phi}{\partial x} + \hat{j}\frac{\partial\phi}{\partial y} + \hat{k}\frac{\partial\phi}{\partial z}$$

Now, comparing the coefficients of $\hat{i}$, $\hat{j}$ and $\hat{k}$ on both sides, we have

$$\frac{\partial \phi}{\partial x} = 2xyz, \ \frac{\partial \phi}{\partial y} = x^2 z + 2y, \ \frac{\partial \phi}{\partial z} = x^2 y$$

Also,

$$\begin{aligned} d\phi &= \frac{\partial \phi}{\partial x} dx + \frac{\partial \phi}{\partial y} dy + \frac{\partial \phi}{\partial z} dz \\ &= 2xyzdx + \left(x^2 z + 2y\right) dy + x^2 ydz \end{aligned}$$

Integrating both sides,

$$\int d\phi = \int_{\substack{y,\,z \\ \text{constant}}} 2xyzdx + \int_{\substack{x,\,z \\ \text{constant}}} (x^2 z + 2y)dy + \int_{\substack{x,\,y \\ \text{constant}}} (x^2 y)dz$$

Considering only those terms on the right-hand side of the integral which have not appeared in the previous integral, i.e., omitting the x^2yz terms in second and third integral, we get

$$\phi = x^2 yz + y^2 + c$$

where c is the integrating constant.

(ii) $\overrightarrow{F}$ is conservation and hence the work done is independent of the path.

$$\begin{aligned} \therefore \text{Work done} &= \int_C \overrightarrow{F} \cdot d\overrightarrow{r} \\ &= \int_{(0,1,1)}^{(1,2,0)} d\phi = |\phi|_{(0,1,1)}^{(1,2,0)} \\ &= \left|x^2 yz + y^2 + c\right|_{(0,1,1)}^{(1,2,0)} \\ &= 3 \end{aligned}$$

Illustration 4.7 If $\overrightarrow{F} = (x^2 - yz)\hat{i} + \left(y^2 - zx\right)\hat{j} + \left(z^2 - xy\right)\hat{k}$, then

(i) If $\overrightarrow{F}$ is conservative, then find its scalar potential ϕ.

(ii) Find the work done in moving a particle under this force field from $(1, 1, 0)$ to $(2, 0, 1)$.

Solution:

(i) Since $\overrightarrow{F}$ is conservative, we have $\overrightarrow{F} = \nabla\phi$

$$\therefore (x^2 - yz)\hat{i} + \left(y^2 - zx\right)\hat{j} + \left(z^2 - xy\right)\hat{k} = \hat{i}\frac{\partial\phi}{\partial x} + \hat{j}\frac{\partial\phi}{\partial y} + \hat{k}\frac{\partial\phi}{\partial z}$$

Now, comparing the coefficients of $\hat{i},\ \hat{j}$ and $\hat{k}$ on both sides, we have

$$\frac{\partial\phi}{\partial x} = (x^2 - yz),\ \frac{\partial\phi}{\partial y} = \left(y^2 - zx\right),\ \frac{\partial\phi}{\partial z} = \left(z^2 - xy\right)$$

Also,

$$\begin{aligned} d\phi &= \frac{\partial\phi}{\partial x}dx + \frac{\partial\phi}{\partial y}dy + \frac{\partial\phi}{\partial z}dz \\ &= (x^2 - yz)dx + \left(y^2 - zx\right)dy + \left(z^2 - xy\right)dz \end{aligned}$$

Integrating both sides,

$$\int d\phi = \int\limits_{\substack{y,\,z \\ constant}} 2xyzdx + \int\limits_{\substack{x,\,z \\ constant}} (x^2z + 2y)dy + \int\limits_{\substack{x,\,y \\ constant}} (x^2y)dz$$

Considering only those terms on the right-hand side of the integral which have not appeared in the previous integral, i.e., omitting the xyz terms in second and third integral, we get

$$\phi = \frac{x^3}{3} - xyz + \frac{y^3}{3} + \frac{z^3}{3} + c$$

where c is the integrating constant.

(ii) $\overrightarrow{F}$ is conservation and hence the work done is independent of the path.

$$\therefore \text{Work done} = \int_C \vec{F} \cdot d\vec{r}$$

$$= \int_{(1,1,0)}^{(2,0,1)} d\phi = |\phi|_{(1,1,0)}^{(2,0,1)}$$

$$= \left| \frac{x^3}{3} - xyz + \frac{y^3}{3} + \frac{z^3}{3} + c \right|_{(1,1,0)}^{(2,0,1)}$$

$$= \frac{7}{3}$$

4.4 Surface Integrals

The surface integral over a curved surface S is the generalization of a double integral over a plane region R given in Figure 4.5.

Let $\vec{F} = F_1\hat{i} + F_2\hat{j} + F_3\hat{k}$ be a continuous vector point function defined over a two-sided surface S. Divide S into a finite number of subsurfaces $S_1, S_2, S_3, \ldots, S_m$ with surface areas $\delta S_1, \delta S_2, \delta S_3, \ldots, \delta S_m$. Let δS_r be the surface area of S_r and $\hat{n}_r$ be the limit vector at some point P_r (in S_r) in the direction of the outward normal to S_r. If we increase the number of sub-surfaces, then the surface area δS_r of each sub-surface will decrease. Thus, as $m \to \infty$, $\delta S_r \to 0$ then,

$$\lim_{m\to\infty} \sum_{r=1}^{m} \vec{F}(P_r) \cdot \hat{n}_r \, \delta S_r = \iint_S \vec{F} \cdot \hat{n} \, dS$$

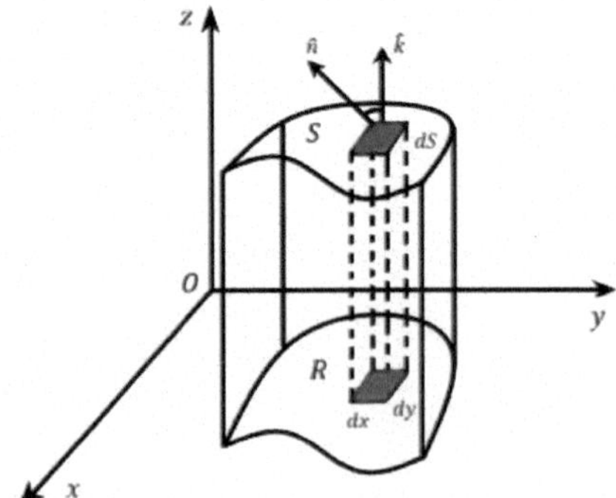

Figure 4.5 Representation of curved surface S and a plane region R

This is called the surface integral of $\overrightarrow{F}$ over the surface S.

The surface integral can also be written as

$$\iint_S \overrightarrow{F} \cdot d\overrightarrow{S}, \text{ Where } d\overrightarrow{S} = \hat{n}dS$$

If the equation of surface S is $\phi(x, y, z) = 0$, then $\hat{n} = \frac{\nabla\phi}{|\nabla\phi|}$.

4.4.1 Flux

If $\overrightarrow{F}$ represents the velocity of the fluid at any point P on a closed surface S, then surface integral $\iint_S \overrightarrow{F} \cdot \hat{n}\, dS$ represents the flux of $\overrightarrow{F}$ over S, i.e., the volume of the fluid flowing out from S per unit time.

Note:

If $\iint_S \overrightarrow{F} \cdot \hat{n}\, dS = 0$, then $\overrightarrow{F}$ is called a solenoidal vector point function.

4.4.2 Evaluation of Surface Integral

A surface integral is evaluated by expressing it as a double integral over the region R. The region R is the orthogonal projection of S on one of the coordinate planes $(xy,\ yz \text{ or } zx)$. Let R be the orthogonal projection of S on the $xy-$plane and $\cos\alpha, \cos\beta, \cos\gamma$ are the direction cosines of $\hat{n}$.

Then

$$\hat{n} = \cos\alpha\ \hat{i} + \cos\beta\ \hat{j} + \cos\gamma\ \hat{k}$$

$$dxdy = \text{Projection of } dS \text{ on } xy - \text{plane} = dS\cos\gamma$$

$$dS = \frac{dxdy}{\cos\gamma} = \frac{dxdy}{\left|\hat{n} \cdot \hat{k}\right|}$$

Hence,

$$\iint_S \overrightarrow{F} \cdot \hat{n}\, dS = \iint_R \overrightarrow{F} \cdot \hat{n}\frac{dxdy}{\left|\hat{n} \cdot \hat{k}\right|}$$

Similarly, taking projection on yz and zx-plane,

$$\iint_S \overrightarrow{F} \cdot \hat{n}\, dS = \iint_R \overrightarrow{F} \cdot \hat{n}\frac{dydz}{\left|\hat{n} \cdot \hat{i}\right|}$$

and

$$\iint_S \overrightarrow{F} \cdot \hat{n}\, dS = \iint_R \overrightarrow{F} \cdot \hat{n}\frac{dzdx}{\left|\hat{n} \cdot \hat{j}\right|}$$

4.4.2.1 Component form of Surface Integral

$$\iint_S \vec{F} \cdot \hat{n}\, dS = \iint_S (F_1\hat{i} + F_2\hat{j} + F_3\hat{k}) \cdot \left(\cos\alpha\, \hat{i} + \cos\beta\, \hat{j} + \cos\gamma\, \hat{k}\right) dS$$
$$= \iint_S F_1 dS\cos\alpha + F_2 dS\cos\beta + F_3 dS\cos\gamma$$
$$= \iint_S F_1 dydz + F_2 dzdx + F_3 dxdy$$

Illustration 4.8: Evaluate $\iint_S \vec{F} \cdot \hat{n}\, dS$, where $\vec{F} = 18z\hat{i} - 12\hat{j} + 3y\hat{k}$ and S is the part of the plane $2x + 3y + 6z = 12$ in the first octant.

Solution:

(i) The given surface is the plane $2x + 3y + 6z = 12$ in the first octant (See Figure 4.6(a)).

Let $\phi = 2x + 3y + 6z$

$$\therefore \hat{n} = \frac{\nabla\phi}{|\nabla\phi|} = \frac{2\hat{i} + 3\hat{j} + 6\hat{k}}{\sqrt{4 + 9 + 36}} = \frac{2\hat{i} + 3\hat{j} + 6\hat{k}}{7}$$

(ii) Let R be the projection of the plane $2x + 3y + 4z = 12$ (in the first octant) on the xy-plane, which is a triangle OAB bounded by the lines $y = 0$, $x = 0$ and $2x + 3y = 12$ (See Figure 4.6(b)).

(iii) Next, we find

$$dS = \frac{dxdy}{\left|\hat{n} \cdot \hat{k}\right|} = \frac{7}{6} dxdy$$

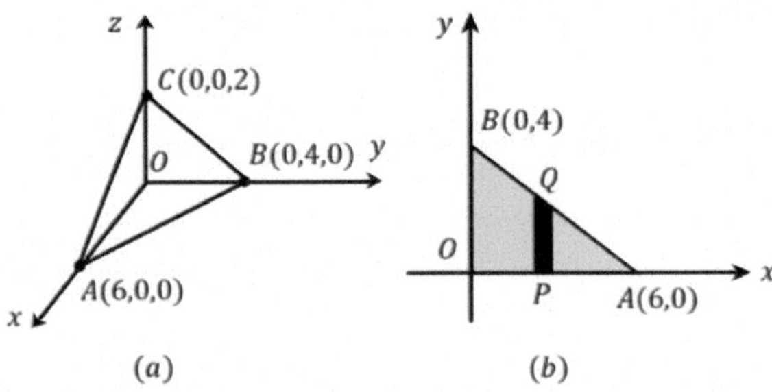

Figure 4.6 Representation of the projection of the plane in the first octant

(iv) Along the vertical strip PQ, y varies from 0 to $\frac{12-2x}{3}$, and in the region R, x varies from 0 to 6.

$$\therefore \iint_S \overrightarrow{F} \cdot \hat{n}\, dS$$

$$= \iint_R \left(18z\hat{\imath} - 12\hat{\jmath} + 3y\hat{k}\right) \cdot \left(\frac{2\hat{\imath} + 3\hat{\jmath} + 6\hat{k}}{7}\right) \frac{7}{6}\, dxdy$$

$$= \frac{1}{6} \iint_R (36z - 36 + 18y)\, dxdy$$

$$= 3 \iint_R \left[2\left(\frac{12 - 2x - 3y}{6}\right) - 2 + y\right] dxdy$$

$$= \int_0^6 \int_0^{\frac{12-2x-3y}{6}} (6 - 2x) dydx$$

$$= 2 \int_0^6 (3 - x)\, |y|_0^{\frac{12-2x}{3}}\, dx$$

$$= 2 \int_0^6 (3 - x) \frac{(12 - 2x)}{3} dx$$

$$= \frac{4}{3} \int_0^6 (x^2 - 9x + 18) dx$$

$$= \frac{4}{3} \left| \frac{x^3}{3} - \frac{9x^2}{2} + 18x \right|_0^6$$

$$= \frac{4}{3}(72 - 162 + 108) = 24$$

Illustration 4.9: Evaluate $\iint_S (yzdydz + xzdzdx + xydxdy)$ over the surface of the sphere $x^2 + y^2 + z^2 = 1$ in the positive octant.

Solution:

(i) $\iint_S \overrightarrow{F} \cdot \hat{n}\, dS = yzdydz + xzdzdx + xydxdy$

$$\therefore \overrightarrow{F} = yz\hat{\imath} + xz\hat{\jmath} + xy\hat{k}$$

(ii) The given surface is the sphere $x^2 + y^2 + z^2 = 1$.

Let $\phi = x^2 + y^2 + z^2$

$$\therefore \hat{n} = \frac{\nabla\phi}{|\nabla\phi|} = \frac{2\text{x}\hat{\imath} + 2y\hat{\jmath} + 2z\hat{k}}{\sqrt{4x^2 + 4y^2 + 4z^2}} = x\hat{\imath} + y\hat{\jmath} + z\hat{k}$$

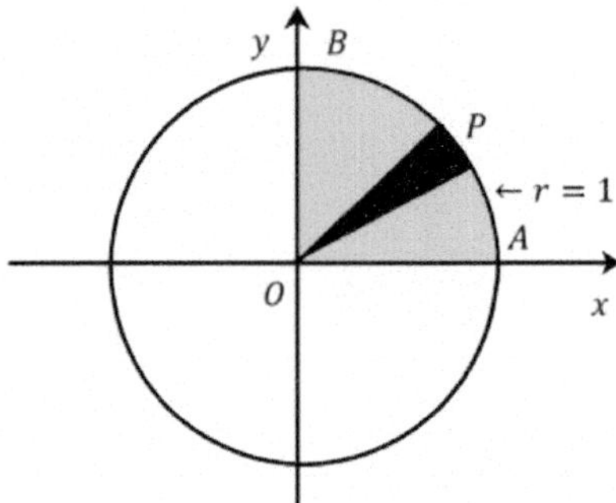

Figure 4.7 Representation of the positive octant of the sphere

$$(\because\ x^2 + y^2 + z^2 = 1)$$

(iii) Let R be the projection of the sphere $x^2 + y^2 + z^2 = 1$ (in the positive octant) on the $xy-$plane $(z = 0)$, which is the part of the circle $x^2+y^2 = 1$ in the first quadrant (See Figure 4.7).

(iv) $dS = \frac{dxdy}{|\hat{n}\cdot\hat{k}|} = \frac{dxdy}{z}$

(v) $\iint_S \overrightarrow{F} \cdot \hat{n}\, dS$

$$= \iint_S yzdydz + xzdzdx + xydxdy$$
$$= \iint_R \left(yz\hat{i} + xz\hat{j} + xy\hat{k}\right) \cdot \left(x\hat{i} + y\hat{j} + z\hat{k}\right) \frac{dxdy}{z}$$
$$= \iint_R (3xyz)\frac{dxdy}{z}$$
$$= 3\iint_R xydxdy$$

Substituting $x = r\cos\theta$, $y = r\sin\theta$, the equation of the circle $x^2+y^2 = 1$ reduces to $r = 1$ and $dxdy = rdrd\theta$.

Along the radius vector OP, r varies from 0 to 1, and in the first quadrant of the circle, θ varies from 0 to $\frac{\pi}{2}$.

$$\therefore \iint_S yzdydz + xzdzdx + xydxdy$$

$$= 3\int_0^{\frac{\pi}{2}}\int_0^1 r\cos\theta \cdot r\sin\theta \cdot r\, dr d\theta$$

$$= 3\int_0^{\frac{\pi}{2}} \frac{\sin 2\theta}{2} d\theta \cdot \int_0^1 r^3 dr$$

$$= \frac{3}{2}\left|\frac{-\cos 2\theta}{2}\right|_0^{\frac{\pi}{2}} \cdot \left|\frac{r^4}{4}\right|_0^1 = \frac{3}{16}(-\cos\pi + \cos 0\ ) = \frac{3}{8}$$

Illustration 4.10: Find the flux of $\overrightarrow{F} = \hat{\imath} - \hat{\jmath} + xyz\hat{k}$ through the circular region S obtained by cutting the sphere $x^2+y^2+z^2 = a^2$ with a plane $y = x$.

Solution:
We know that

$$\text{Flux} = \iint_S \overrightarrow{F} \cdot \hat{n}\, dS$$

(i) Here, the surface S (See Figure 4.8) is the intersection of the sphere $x^2+y^2+z^2 = a^2$ with a plane $y = x$, which is an ellipse $2x^2+z^2 = a^2$.

(ii) Normal to the ellipse $2x^2 + z^2 = a^2$ is also normal to the plane $y = x$.

Let $\phi = x - y$

$$\hat{n} = \frac{\nabla\phi}{|\nabla\phi|} = \frac{\hat{\imath} - \hat{\jmath}}{\sqrt{2}}$$

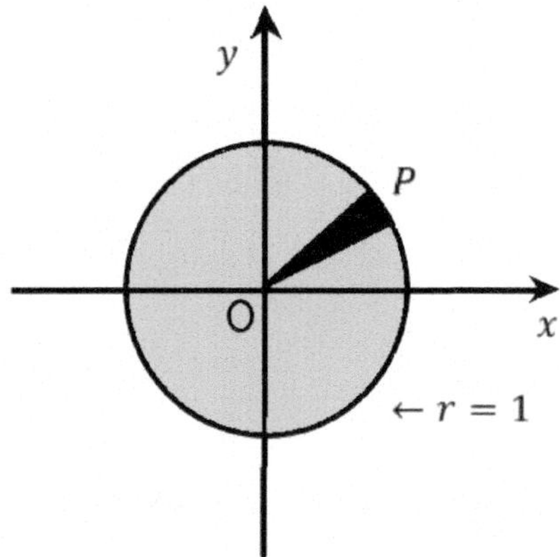

Figure 4.8 Representation of the sphere

(iii) Let R be the projection of the surface S on the xz-plane, which is an ellipse $2x^2 + z^2 = a^2$

(iv) $dS = \frac{dxdz}{|\hat{n}\cdot\hat{j}|} = \sqrt{2}\, dxdz$

(v) $\iint_S \vec{F}\cdot\vec{n}\, dS = \iint_R \left(\hat{i} - \hat{j} + xyz\hat{k}\right)\cdot\left(\frac{\hat{i}-\hat{j}}{\sqrt{2}}\right)\sqrt{2}\, dxdz$

$$= \iint_R 2dxdz$$

Substituting $x = \frac{a}{\sqrt{2}} r\cos\theta$, $z = ar\sin\theta$, the equation of the ellipse $2x^2 + z^2 = a^2$ reduces to $r = 1$ and $dxdz = \frac{a^2}{\sqrt{2}} rdr\, d\theta$.

Along the radius vector OP, r varies from 0 to 1, and for a complete ellipse, θ varies from 0 to 2π.

$$\begin{aligned}\iint_S \vec{F}\cdot\vec{n}\, dS &= 2\int_0^{2\pi}\int_0^1 \frac{a^2}{\sqrt{2}} rdr\, d\theta \\ &= \frac{2a^2}{\sqrt{2}}\left|\frac{r^2}{2}\right|_0^1 |\theta|_0^{2\pi} = \sqrt{2}\, a^2\cdot\frac{1}{2}\cdot 2\pi \\ &= \sqrt{2}\,\pi a^2\end{aligned}$$

Aliter,

$$\begin{aligned}\iint_S \vec{F}\cdot\vec{n}\, dS &= 2\iint_R dxdz \\ &= 2\left[Area\ of\ the\ ellipse\ \frac{x^2}{\left(\frac{a}{\sqrt{2}}\right)^2} + \frac{y^2}{a^2} = 1\right] \\ &= 2\cdot\pi\frac{a}{\sqrt{2}}\cdot a \\ &= \sqrt{2}\,\pi a^2\end{aligned}$$

Hence, flux $= \sqrt{2}\,\pi a^2$.

4.5 Volume Integrals

If V be a region in space bounded by a closed surface S, then the volume integral of a vector point function $\vec{F}$ is $\iiint_V \vec{F}\, dV$.

4.5.1 Component Form of Volume Integral

If $\vec{F} = F_1\hat{i} + F_2\hat{j} + F_3\hat{k}$ then

$$\iiint_V \vec{F}\, dV = \iiint_V \left(F_1\hat{i} + F_2\hat{j} + F_3\hat{k}\right) dxdydz$$

$$= \hat{i}\iiint F_1 dxdydz + \hat{j}\iiint F_2 dxdydz + \hat{k}\iiint F_3 dxdydz$$

Another type of volume integral is $\iiint_V \phi\, dV$, where ϕ is a scalar function.

Illustration 4.11: Evaluate $\iiint_V \vec{F}\, dV$ where $\vec{F} = x\hat{i} + y\hat{j} + 2z\hat{k}$ and V is the volume enclosed by the planes $x = 0$, $y = a$, $z = b^2$ and the surface $z = x^2$.

Solution:

(i) V is the volume of the cylinder in positive octant with the base as OAB and bounded between the planes $y = 0$ and $y = a$. y varies from 0 to a (See Figure 4.9).

(ii) Along the vertical strip PQ, z varies from x^2 to b^2, and in the region OAB, x varies from 0 to b.

$$\iiint_V \vec{F} dV = \int_{x=0}^{b}\int_{z=x^2}^{b^2}\int_{y=0}^{a} \left(x\hat{i} + y\hat{j} + 2z\hat{k}\right) dxdydz$$

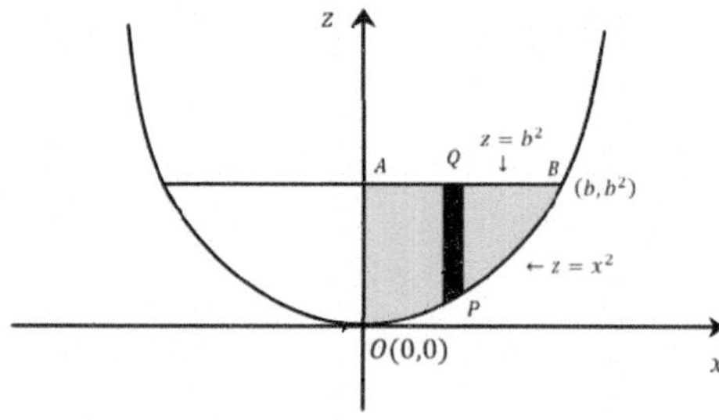

Figure 4.9 Representation of the cylinder in positive octant

$$= \int_0^b \int_{x^2}^{b^2} \left(x\hat{\imath}|y|_0^a + \hat{\jmath}\left|\frac{y^2}{2}\right|_0^a + 2z\hat{k}|y|_0^a \right) dzdx$$

$$= \int_0^b \int_{x^2}^{b^2} \left(\hat{\imath}xa + \hat{\jmath}\frac{a^2}{2} + \hat{k}2za \right) dzdx$$

$$= \int_0^b \left(\hat{\imath}xa|z|_{x^2}^{b^2} + \hat{\jmath}\frac{a^2}{2}|z|_{x^2}^{b^2} + \hat{k}a|z^2|_{x^2}^{b^2} \right) dx$$

$$= \int_0^b \left(\hat{\imath}xa(b^2 - x^2) + \hat{\jmath}\frac{a^2}{2}(b^2 - x^2) + \hat{k}a(b^4 - x^4) \right) dx$$

$$= \left| \hat{\imath}a\left(\frac{b^2x^2}{2} - \frac{x^4}{4}\right) + \hat{\jmath}\frac{a^2}{2}\left(b^2x - \frac{x^3}{3}\right) + \hat{k}a\left(b^4x - \frac{x^5}{5}\right) \right|_0^b$$

$$= \hat{\imath}a\left(\frac{b^4}{2} - \frac{b^4}{4}\right) + \hat{\jmath}\frac{a^2}{2}\left(b^3 - \frac{b^3}{3}\right) + \hat{k}a\left(b^5 - \frac{b^5}{5}\right)$$

$$= a\frac{b^4}{4}\hat{\imath} + \frac{a^2b^3}{3}\hat{\jmath} + \frac{4ab^5}{5}\hat{k}.$$

Illustration 4.12: Evaluate $\iiint_V \left(\nabla \times \vec{F}\right) dV$, where $\vec{F} = (2x^2 - 3z)\,\hat{\imath} - 2xy\hat{\jmath} - 4x\hat{k}$ and V is the closed region bounded by the planes $x = 0$, $y = 0$, $z = 0$ and $2x + 2y + z = 4$.

Solution:

(i)

$$\nabla \times \vec{F} = \begin{vmatrix} \hat{\imath} & \hat{\jmath} & \hat{k} \\ \frac{\partial}{\partial x} & \frac{\partial}{\partial y} & \frac{\partial}{\partial z} \\ 2x^2 - 3z & -2xy & -4x \end{vmatrix}$$

$$= (0 - 0)\,\hat{\imath} - (-4 + 3)\,\hat{\jmath} + (-2y - 0)\,\hat{k}$$

$$= \hat{\jmath} - 2y\hat{k}$$

(ii) Along the elementary volume PQ, z varies from 0 to $4 - 2x - 2y$. Along the vertical strip $P'Q'$, y varies from 0 to $2 - x$, and in the region, x varies from 0 to 2.

$$\iiint_V \left(\nabla \times \vec{F}\right) dV = \int_0^2 \int_0^{2-x} \int_0^{4-2x-2y} \left(\hat{\jmath} - 2y\hat{k}\right) dxdydz$$

$$= \int_0^2 \int_0^{2-x} \left(\hat{\jmath} - 2y\hat{k}\right) |z|_0^{4-2x-2y} dydx$$

$$= \int_0^2 \int_0^{2-x} \left(\hat{\jmath} - 2y\hat{k}\right) (4 - 2x - 2y)\, dydx$$

$$= \int_0^2 \int_0^{2-x} \left[(4 - 2x - 2y)\,\hat{\jmath} - 2\,(4 - 2x)\, y\hat{k} + 4y^2\hat{k}\right] dydx$$

$$= \int_0^2 \left[\left\{(4 - 2x)\, |y|_0^{2-x} - \left|y^2\right|_0^{2-x}\right\} \hat{\jmath}\right.$$

$$\left. - \left\{2\,(2 - x)\, \left|y^2\right|_0^{2-x} - 4\left|\frac{y^3}{3}\right|_0^{2-x}\right\} \hat{k}\right] dx$$

$$= \int_0^2 \left[(2 - x)^2\hat{\jmath} - \frac{2}{3}(2 - x)^3\hat{k}\right] dx = \frac{8}{3}\hat{\jmath} - \frac{8}{3}\hat{k} = \frac{8}{3}(\hat{\jmath} - \hat{k})$$

4.6 Exercise

1. Evaluate $\int_C \overrightarrow{F} \cdot d\overrightarrow{r}$ along the curve $x^2 + y^2 = 1,\ z = 1$ in the positive direction from $(0, 1, 1)$ to $(1, 0, 1)$, where $\overrightarrow{F} = (yz + 2x)\,\hat{\imath} + xz\hat{\jmath} + (xy + 2z)\,\hat{k}$.

   (Answer: 1)

2. Evaluate $\int_C \overrightarrow{F} \cdot d\overrightarrow{r}$ over the circular path $x^2 + y^2 = a^2$ where $\overrightarrow{F} = \sin y\ \hat{\imath} + x\,(1 + \cos y\,)\,\hat{\jmath}$.

   (Answer : πa^2)

3. Find the work done in moving a particle in the force field $\overrightarrow{F} = 3x^2\hat{\imath} + (2xz - y)\,\hat{\jmath} + z\hat{k}$ along the curve $x^2 = 4y$ and $3x^3 = 8z$ from $x = 0$ to $x = 2$.

   (Answer: 16)

4. Find the work done in moving a particle from $A(1, 0, 1)$ to $B(2, 1, 2)$ along the straight line AB in the force field $\overrightarrow{F} = x^2\hat{\imath} + (x - y)\,\hat{\jmath} + (y + z)\,\hat{k}$.

   $\left(\text{Answer : } \frac{16}{3}\right)$

5. Find the work done in moving a particle along the straight-line segments joining the points $(0,0,0)$ to $(1,0,0)$, then to $(1,1,0)$, and finally to $(1,1,1)$ under the force field $\overrightarrow{F} = (3x^2+6y)\,\hat{i} - 14yz\hat{j} + 20xz^2\hat{k}$.

   $\left(\text{Answer : } \frac{23}{3}\right)$

6. Find the work done by the force $\overrightarrow{F} = x\hat{i} - z\hat{j} + 2y\hat{k}$ in displacing the particle along the triangle OAB, where

$$
\begin{aligned}
OA &: 0 \le x \le 1, \quad y = x, \quad z = 0 \\
AB &: 0 \le z \le 1, \quad x = 1, \quad y = 1 \\
BO &: 0 \le x \le 1, \quad y = z = x
\end{aligned}
$$

   $\left(\text{Answer : } \frac{3}{2}\right)$

7. Find the work done by the force $\overrightarrow{F} = 16y\hat{i} + (3x^2+2)\,\hat{j}$ in moving a particle once round the right half of the ellipse $x^2 + a^2y^2 = a^2$ from $(0,1)$ to $(0,-1)$.

   $\left(\text{Answer : } 8a\pi - 4a^2 - 4\right)$

8. Evaluate $\int_C \overrightarrow{F} \cdot d\overrightarrow{r}$, where $\overrightarrow{F} = 2x\hat{i} + 4y\hat{j} - 3z\hat{k}$ and C is the curve $\overrightarrow{r} = \cos t\ \hat{i} + \sin t\ \hat{j} + t\ \hat{k}$ from $t = 0$ to $t = \pi$.

   $\left(\text{Answer :} \frac{-3\pi^2}{2}\right)$

9. Find the circulation of $\overrightarrow{F} = (x-3y)\,\hat{i} + (y-2x)\,\hat{j}$ around the ellipse in the $xy-$ plane with the origin as a center and 2 and 3 as semi-major and semi-minor axes respectively.

   $\left(\text{Answer : } 6\pi\right)$

10. Find the circulation of $\overrightarrow{F} = y\hat{i} + z\hat{j} + x\hat{k}$ around the curve $x^2 + y^2 = 1,\ z = 0$.

    $\left(\text{Answer : } -\pi\right)$

11. If $\overrightarrow{F} = (2xy + z^3)\,\hat{i} + x^2\hat{j} + 3xz^2\hat{k}$ is conservative then (i) find its scalar potential ϕ, (ii) find the work done in moving a particle under this force field from $(1,-2,1)$ to $(3,1,4)$.

    $\left(\text{Answer : } (i)\,\phi = x^2y + xz^3 + c,\ (ii)\,202\right)$

12. If $\overrightarrow{F} = 3x^2y\hat{i} + (x^3 - 2yz^2)\hat{j} + (3z^2 - 2y^2z)\hat{k}$ is conservative then (i) find its scalar potential ϕ, (ii) find the work done in moving a particle under this force field from $(2, 1, 1)$ to $(2, 0, 1)$.

$$\left(\text{Answer}: (i)\, \phi = x^3y + z^3 - y^2z^2 + c,\ (ii) - 7\right)$$

13. If $\overrightarrow{F} = 2xye^z\hat{i} + x^2e^z\hat{j} + x^2ye^z\hat{k}$ is conservative then (i) find its scalar potential ϕ, (ii) find the work done in moving a particle under this force field from $(0, 0, 0)$ to $(1, 1, 1)$.

$$\left(\text{Answer}: (i)\, \phi = x^2ye^z + c,\ (ii)\, e\right)$$

14. Evaluate $\iint_S \overrightarrow{F} \cdot \hat{n}dS$ where $\overrightarrow{F} = 3y\hat{i} + 2z\hat{j} + x^2yz\hat{k}$ and S is the surface $y^2 = 5x$ in the positive octant bounded by the planes $x = 3$ and $z = 4$.

$$(\text{Answer}: -42)$$

15. Evaluate $\iint_S \overrightarrow{F} \cdot \hat{n}dS$ where $\overrightarrow{F} = (x + y^2)\hat{i} - 2x\hat{j} + 2yz\hat{k}$ and S is the surface $2x + y + 2z = 6$ in the first octant.

$$(\text{Answer}: 81)$$

16. Evaluate $\iint_S \nabla \times \overrightarrow{F} \cdot \hat{n}dS$ where $\overrightarrow{F} = y^2\hat{i} + y\hat{j} - xz\hat{k}$ and S is the upper half of the sphere $x^2 + y^2 + z^2 = a^2$.

$$(\text{Answer}: 0)$$

17. Find the flux of the vector field $\overrightarrow{F}$ through the portion of the sphere $x^2 + y^2 + z^2 = 36$ lying between the planes $z = \sqrt{11}$ and $z = \sqrt{20}$ where $\overrightarrow{F} = x\hat{i} + y\hat{j} + z\hat{k}$.

$$\left(\text{Answer}: 72\pi\sqrt{20} - \sqrt{11}\right)$$

18. Find the flux of the vector field $\overrightarrow{F} = x\hat{i} + y\hat{j} + \sqrt{x^2 + y^2 - 1}\,\hat{k}$ through the outer side of the hyperboloid $z = \sqrt{x^2 + y^2 - 1}$ bounded by the planes $z = 0$ and $z = \sqrt{3}$.

$$\left(\text{Answer}: 2\sqrt{3}\pi\right)$$

19. Find the flux of the vector field $\overrightarrow{F} = 2y\hat{i} - z\hat{j} + x^2\hat{k}$ across the surface of the parabolic cylinder $y^2 = 8x$ in the first octant bounded by the planes $y = 4$ and $z = 6$.

$$(\text{Answer}: 132)$$

20. Evaluate $\iiint_V \left(\nabla \cdot \vec{F}\right) dV$ where $\vec{F} = 2x^2y\hat{\imath} - y^2\hat{\jmath} + 4xz^2\hat{k}$ and $V$is the region in the first octant bounded by the cylinder $y^2 + z^2 = 9$ and the plane $z = 2$.

(Answer : 180)

21. Evaluate $\iiint_V \left(\nabla \cdot \vec{F}\right) dV$ where $\vec{F} = 2xz\hat{\imath} - x\hat{\jmath} + y^2\hat{k}$ and V is the region bounded by the surfaces $x = 0,\ y = 0, y = 6, z = x^2, z = 4$.

$\left(\text{Answer} : 128\hat{\imath} - 24\hat{\jmath} + 384\hat{k}\right)$

22. Evaluate $\iiint_V f dV$ where $f = 45x^2y$ and V is the region bounded by the planes $4x + 2y + z = 8,\ x = 0, y = 0, z = 0$.

(Answer : 128)

23. Evaluate $\iiint_V \left(\nabla \times \vec{F}\right) dV$, where $\vec{F} = (x + 2y)\hat{\imath} - 3z\hat{\jmath} + x\hat{k}$ and V is the closed region in the first octant bounded by the plane $2x + 2y + z = 4$.

$\left(\text{Answer} : \frac{8}{3}\left(3\hat{\imath} - \hat{\jmath} + 2\hat{k}\right)\right)$

5

Green's Theorem, Stokes' Theorem, and Gauss' Theorem

5.1 Green's Theorem (in the Plane)

Double integrals over a plane region could also be transformed into line integrals over the boundary of the region and conversely. This is of practical interest because it's going to help to form the evaluation of an integral easier. The transformation can be computed using a theorem known as Green's theorem.

Statement: If $M(x, y)$, $N(x, y)$ and their partial derivatives $\frac{\partial M}{\partial y}$, $\frac{\partial N}{\partial x}$ are continuous in some region R of xy-plane bounded by a closed curve C, then

$$\oint_C (Mdx + Ndy) = \iint_R \left(\frac{\partial N}{\partial x} - \frac{\partial M}{\partial y} \right) dxdy$$

Proof: Let R be the bounded region by the curve C as shown in Figure 5.1. Let the curve C be divided into two parts, the curves EAB and BDE.

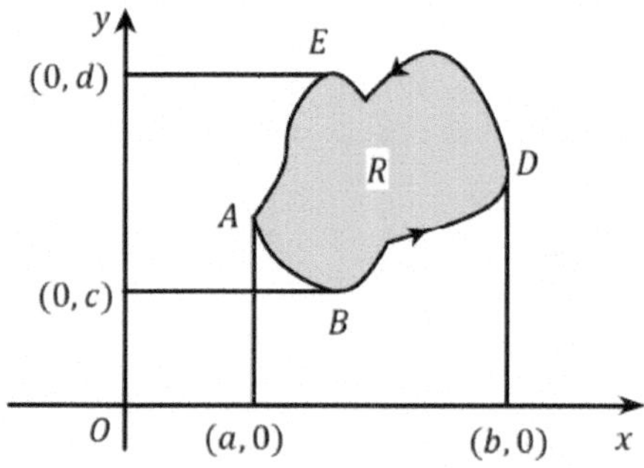

Figure 5.1 Represents the region R bounded by the curve C

Let the equations of the curves EAB and BDE are $x = f_1(y)$, $x = f_2(y)$ respectively and are bounded between the lines $y = c$ and $y = d$.
Consider,

$$\iint_R \frac{\partial N}{\partial x} dxdy = \int_c^d \left[\int_{f_1(y)}^{f_2(y)} \frac{\partial N}{\partial x} dx \right] dy$$

$$= \int_c^d |N(x,y)|_{f_1(y)}^{f_2(y)} dy$$

$$= \int_c^d [N(f_2, y) - N(f_1, y)] \, dy$$

$$= \int_c^d N(f_2, y) \, dy + \int_c^d N(f_1, y) \, dy$$

$$= \int_{BDE} N(x,y) \, dy + \int_{EAB} N(x,y) \, dy$$

$$= \oint_C N(x,y) dy$$

$$\therefore \iint_R \frac{\partial N}{\partial x} dxdy = \oint_C N(x,y) dy \tag{5.1}$$

Similarly, let the curve C be divided into two parts, the curves ABD and DEA. Let the equation of the curves ABD and DEA are $y = g_1(x)$, $y = g_2(x)$ respectively and are bounded between the lines $x = a$ and $x = b$.
Now, consider

$$\iint_R \frac{\partial M}{\partial y} dxdy = \int_a^b \left[\int_{g_1(x)}^{g_2(x)} \frac{\partial M}{\partial y} dy \right] dx$$

$$= \int_a^b |M(x,y)|_{g_1(x)}^{g_2(x)} dx$$

$$= \int_a^b [M(x, g_2) - M(x, g_1)] \, dx$$

$$= - \int_a^b M(x, g_2) \, dx - \int_a^b M(x, g_1) \, dx$$

$$= - \left[\int_{DEA} M(x,y) \, dx + \int_{ABD} M(x,y) \, dx \right]$$

$$= -\oint_C M(x,y)dx$$

$$\therefore -\iint_R \frac{\partial M}{\partial y} dxdy = \oint_C M(x,y)\,dx \tag{5.2}$$

Adding Equations (5.1) and (5.2), we get

$$\oint_C (M(x,y)\,dx + N(x,y)\,dy) = \iint_R \left(\frac{\partial N}{\partial x} - \frac{\partial M}{\partial y}\right) dxdy$$

or

$$\oint_C (Mdx + Ndy) = \iint_R \left(\frac{\partial N}{\partial x} - \frac{\partial M}{\partial y}\right) dxdy.$$

Note: Let $\overrightarrow{F} = M\hat{i} + N\hat{j}$ and $\overrightarrow{r} = x\hat{i} + y\hat{j}$, then $\overrightarrow{F} \cdot d\overrightarrow{r} = Mdx + Ndy$. Also,

$$\text{curl } \overrightarrow{F} = \begin{vmatrix} \hat{i} & \hat{j} & \hat{k} \\ \frac{\partial}{\partial x} & \frac{\partial}{\partial y} & \frac{\partial}{\partial z} \\ M & N & O \end{vmatrix} = \overrightarrow{k}\left(\frac{\partial N}{\partial x} - \frac{\partial M}{\partial y}\right)$$

The component of the curl $\overrightarrow{F}$ which is normal to a region R in xy-plane is

$$\left(\nabla \times \overrightarrow{F}\right) \cdot \overrightarrow{k} = \frac{\partial N}{\partial x} - \frac{\partial M}{\partial y}$$

Hence, the vector form of Green's theorem is given as

$$\oint_C \overrightarrow{F} \cdot d\overrightarrow{r} = \iint_R \left(\nabla \times \overrightarrow{F}\right) \cdot \overrightarrow{k}\, dxdy$$

where, $\overrightarrow{F} = M\hat{i} + N\hat{j}$, $\overrightarrow{r} = x\hat{i} + y\hat{j}$, $\hat{k}$ is the unit vector along z-axis.

5.1.1 Area of the Plane Region

Let A be the area of the plane region R bounded by a closed curve C.

Let $M = -y, \;\; N = x \Rightarrow \frac{\partial M}{\partial y} = -1, \;\; \frac{\partial N}{\partial x} = 1$

Using Green's theorem,

$$\oint_C (-ydx + xdy) = \iint_R (1+1)dxdy = 2\iint_R dxdy = 2A$$

$$\therefore A = \frac{1}{2}\oint_C (xdy - ydx)$$

Note: In polar coordinates,

$$x = r\cos\theta\ ,\quad y = r\sin\theta$$

$$dx = \cos\theta\ dr - r\sin\theta\ d\theta,\quad dy = \sin\theta\ dr + r\cos\theta\ \ d\theta$$

$$\therefore A = \frac{1}{2}\oint_C \left[r\cos\theta\ (\sin\theta\ dr + r\cos\theta\ d\theta) - r\sin\theta\ (\cos\theta\ dr - r\sin\theta\ d\theta)\right]$$

$$\therefore A = \frac{1}{2}\oint_C r^2 d\theta$$

Illustration 5.1: Using Green's theorem, evaluate

$$\oint_C \left(3x^2 - 8y^2\right) dx + (4y - 6xy)dy$$

where C is the boundary of the region bounded by $x = y^2$ and $y = x^2$.

Solution:
Here, C is the boundary of the region bounded by $x = y^2$ and $y = x^2$ (See Figure 5.2). Therefore, solving both the parabolas, we get

$$\therefore x^4 - x = 0$$
$$\therefore x\left(x^3 - 1\right) = 0$$
$$\therefore x = 0 \text{ or } x^3 = 1$$

Therefore, for $x = 0 \Rightarrow y = 0$ and for $x = 1 \Rightarrow y = 1$. So, both the parabolas are intersecting at $(0, 0)$ and (1,1).

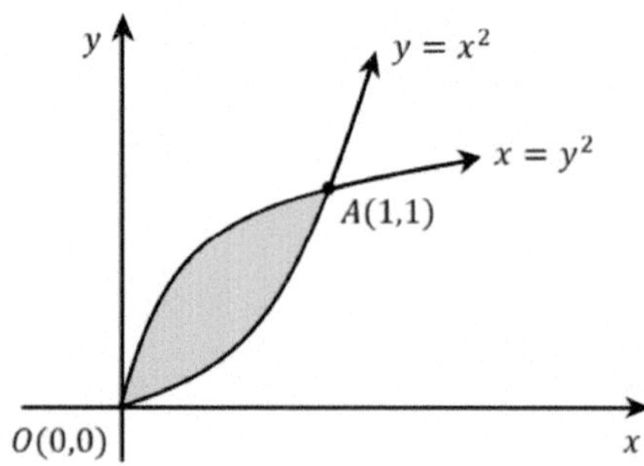

Figure 5.2 Represents the region R bounded two parabolas

Next,

$$M = 3x^2 - 8y^2, \qquad N = 4y - 6xy$$

$$\therefore \frac{\partial M}{\partial y} = -16y, \quad \frac{\partial N}{\partial x} = -6y$$

$$\therefore \oint_C M dx + N dy = \iint_R \left(\frac{\partial N}{\partial x} - \frac{\partial M}{\partial y} \right) dx dy$$

$$= \int_0^1 \int_{x^2}^{\sqrt{x}} 10y \, dx dy$$

$$= \int_0^1 5 \left(y^2 \right)_{x^2}^{\sqrt{x}} dx$$

$$= 5 \int_0^1 \left(x - x^4 \right) dx$$

$$= 5 \left(\frac{x^2}{2} - \frac{x^5}{5} \right)_0^1 = 5 \left(\frac{1}{2} - \frac{1}{5} \right) = \frac{3}{2}$$

Illustration 5.2: Applying Green's theorem, evaluate

$$\oint_C [(y - \sin x \,) \, dx + \cos x \;\, dy]$$

where C is the plane triangle enclosed by the lines $y = 0$, $x = \frac{\pi}{2}$ and $y = \frac{2}{\pi} x$ (See Figure 5.3).

Solution:

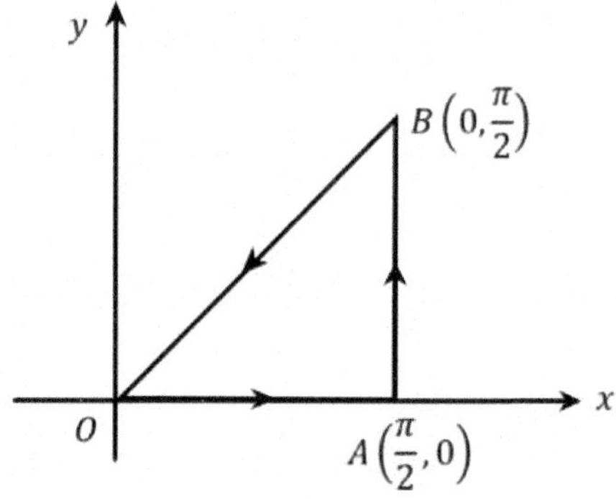

Figure 5.3 Represents the plane triangle enclosed by given lines

Here, $M = y - \sin x$, $N = \cos x$
$\therefore \frac{\partial M}{\partial y} = 1, \frac{\partial N}{\partial x} = -\sin x$ and we know that

$$\oint_C Mdx + Ndy = \iint_R \left(\frac{\partial N}{\partial x} - \frac{\partial M}{\partial y}\right) dxdy$$
$$= \iint_R (-\sin x - 1\ ) \, dydx$$
$$= \int_{x=0}^{\frac{\pi}{2}} \left[\int_{y=0}^{\frac{2x}{\pi}} (-\sin x \ - 1) dy\right] dx$$
$$= \int_0^{\frac{\pi}{2}} [-y\sin x \ - y]_0^{\frac{2x}{\pi}} \, dx$$
$$= \int_0^{\frac{\pi}{2}} \left(-\frac{2\pi}{x}\sin x \ - \frac{2\pi}{x}\right) dx$$
$$= -\frac{2}{\pi}[-x\cos x \ + \sin x\ ]_0^{\frac{\pi}{2}} - \left[\frac{x^2}{\pi}\right]_0^{\frac{\pi}{2}} = -\frac{2}{\pi} - \frac{\pi}{4}$$

Illustration 5.3: Verify Green's theorem for the function $\vec{F} = (x + y)\,\hat{i} + 2xy\hat{j}$ and C is the rectangle in the xy-plane bounded by $x = 0, y = 0, x = a$, $y = b$ (See Figure 5.4).

Solution:

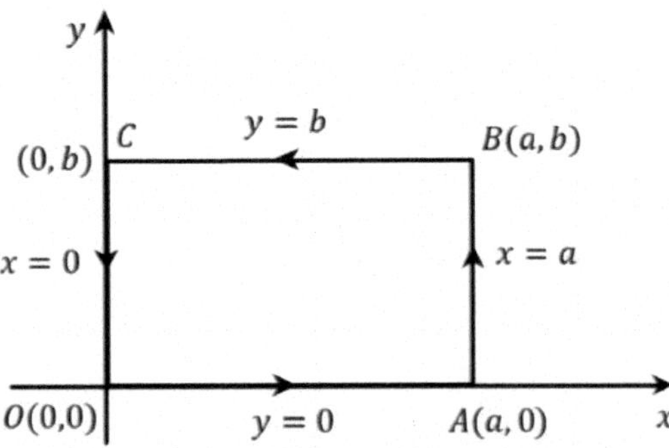

Figure 5.4 Represents the rectangle in the xy-plane bounded by lines

We know that

$$\oint_C Mdx + Ndy = \oint_C \vec{F} \cdot d\vec{r}$$
$$= \oint_C (x+y)\,dx + 2xy\,dy$$

Along OA, $y = 0 \Rightarrow dy = 0$

$$\therefore \int_{OA} Mdx + Ndy = \int_0^a x dx = \frac{a^2}{2} \qquad \text{(i)}$$

Along AB, $x = a \Rightarrow dx = 0$

$$\therefore \int_{AB} Mdx + Ndy = \int_0^b 2ay\,dy = \frac{2ab^2}{2} = ab^2 \qquad \text{(ii)}$$

Along BC, $y = b \Rightarrow dy = 0$

$$\therefore \int_{AB} Mdx + Ndy = \int_a^0 (x+b)dx = -\frac{a^2}{2} - ab \qquad \text{(iii)}$$

Along CO, $x = 0 \Rightarrow dx = 0$.

$$\therefore \int_{CO} Mdx + Ndy = \int_b^0 0\,dy = 0 \qquad \text{(iv)}$$

Adding (i), (ii), (iii), and (iv), we get

$$\oint_C Mdx + Ndy = ab^2 - ab \qquad \text{(v)}$$

Now,

$$\iint_R \left(\frac{\partial N}{\partial x} - \frac{\partial M}{\partial y}\right) dxdy = \int_0^a \int_0^b (2y-1)\,dxdy$$
$$= \int_0^a \left(\frac{2y^2}{2} - y\right)_0^b dx$$
$$= \int_0^a \left(b^2 - b\right) dx$$
$$= \left(b^2 - b\right) \int_0^a dx$$
$$= \left(b^2 - b\right)(x)_0^a = ab^2 - ab \qquad \text{(vi)}$$

From (v) and (vi) we see that

$$\oint_C M dx + N dy = \iint_R \left(\frac{\partial N}{\partial x} - \frac{\partial M}{\partial y} \right) dx dy$$

Hence, Green's theorem is verified.

Illustration 5.4: Use Green's theorem to evaluate the integral

$$\oint_C (y^2 dx + x^2 dy)$$

where, C: The triangle bounded by $x = 0, x + y = 1, y = 0$ (See Figure 5.5).

Solution:
Here, $M = y^2$ and $N = x^2$ then

$$\frac{\partial M}{\partial y} = 2y, \quad \frac{\partial N}{\partial x} = 2x$$

From Green's theorem

$$\oint_C M dx + N dy = \iint_R \left(\frac{\partial N}{\partial x} - \frac{\partial M}{\partial y} \right) dx dy = 2 \int_0^1 \int_0^{1-y} (x - y) dx dy$$

$$= 2 \int_0^1 \left[\frac{x^2}{2} - xy \right]_0^{1-y} dy = 2 \int_0^1 \left[\frac{(1-y)^2}{2} - y(1-y) \right] dy$$

$$= 2 \left[\frac{1}{2} y - y^2 + \frac{1}{2} y^3 \right]_0^1 = 2 \left[\frac{1}{2} - 1 + \frac{1}{2} \right] = 0$$

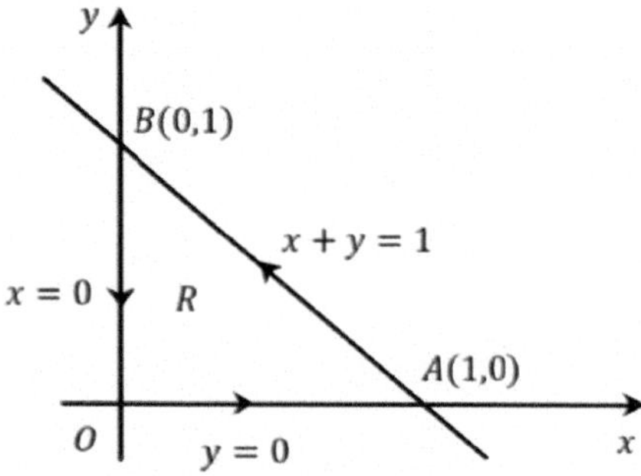

Figure 5.5 Represents the triangle in the xy-plane bounded by lines

Illustration 5.5: Verify Green's theorem for the function $\overrightarrow{F} = (x-y)\,\hat{\imath} + x\hat{\jmath}$ and the region R bounded by the unit circle

$$C\ : \overrightarrow{r}\,(t) = (\text{cos}t\,)\,\hat{\imath} + (\text{sin}t\,)\,\hat{\jmath},\ \ 0 \leq t \leq 2\pi$$

Solution: Consider, $\overrightarrow{r} = x\hat{i} + y\hat{j} \Rightarrow d\overrightarrow{r} = \hat{i}dx + \hat{j}dy$ and $\overrightarrow{F} = (x-y)\hat{i} + x\hat{j}$

$$\therefore \overrightarrow{F} \cdot d\overrightarrow{r} = (x-y)\,dx + xdy$$

Now,

$$M = x - y,\ \ N = x$$

$$\therefore\ \frac{\partial M}{\partial y} = -1,\ \ \frac{\partial N}{\partial x} = 1$$

$$\therefore\ \oint_C Mdx + Ndy = \iint_R \left[\frac{\partial N}{\partial x} - \frac{\partial M}{\partial y}\right] dxdy$$

$$= \iint_R [1+1]\,dxdy$$

$$= 2\iint_R dxdy = 2\pi$$

$$\oint_C \overrightarrow{F} \cdot d\overrightarrow{r} = \oint_C (x-y)\,dx + xdy$$

$$= \oint_0^{2\pi} \left[(\text{cos}t\ - \text{sin}t\,)\,(-\text{sin}t\,) + \text{cos}^2t\,\right] dt$$

$$= \int_0^{2\pi} \left[-\text{sin}t\ \text{cos}t\ + \text{sin}^2t\ + \text{cos}^2t\,\right] dt$$

$$= \int_0^{2\pi} [-\text{sin}t\ \text{cos}t\ + 1]\,dt$$

$$= \int_0^{2\pi} \left[-\frac{\text{sin}2t}{2} + 1\right] dt$$

$$= \frac{1}{2}\left(\frac{\text{cos}2t}{2}\right)_0^{2\pi} + |t|_0^{2\pi}$$

$$= \frac{1}{4}\,[\text{cos}4\pi\ - \text{cos}0\,] + 2\pi = 2\pi$$

$$\therefore\ \oint_C \overrightarrow{F} \cdot d\overrightarrow{r} = \iint_R \left[\frac{\partial N}{\partial x} - \frac{\partial M}{\partial y}\right] dxdy = 2\pi$$

Hence, the theorem is verified.

Illustration 5.6: Verify Green's theorem for $\int_C \left(\frac{1}{y}dx + \frac{1}{x}dy\right)$ where C is the boundary of the region bounded by the parabola $y = \sqrt{x}$ and the lines $x = 1$, $x = 4$, $y = 1$ (See Figure 5.6).

Solution:

(i) The point of intersection of the

(a) parabola $y = \sqrt{x}$ and the line $x = 1$ is obtained as $y = \sqrt{1} = 1$. Hence, $A(1, 1)$ is the point of intersection.

(b) parabola $y = \sqrt{x}$ and line $x = 4$ is obtained as $y = \sqrt{4} = 2$. Hence, $D(4, 2)$ is the point of intersection.

(ii) $M = \frac{1}{y}$, $N = \frac{1}{x}$

$$\frac{\partial M}{\partial y} = -\frac{1}{y^2}, \quad \frac{\partial N}{\partial x} = -\frac{1}{x^2}$$

(iii) $\oint_C (Mdx + Ndy) = \int_{AB} (Mdx + Ndy) + \int_{BD} (Mdx + Ndy)$
$+ \int_{DQA} (Mdx + Ndy)$ (i)

(a) Along $AB : y = 1 \Rightarrow dy = 0$ and x varies from 1 to 4.

$$\int_{AB} (Mdx + Ndy) = \int_{AB} \left(\frac{1}{y}dx + \frac{1}{x}dy\right)$$

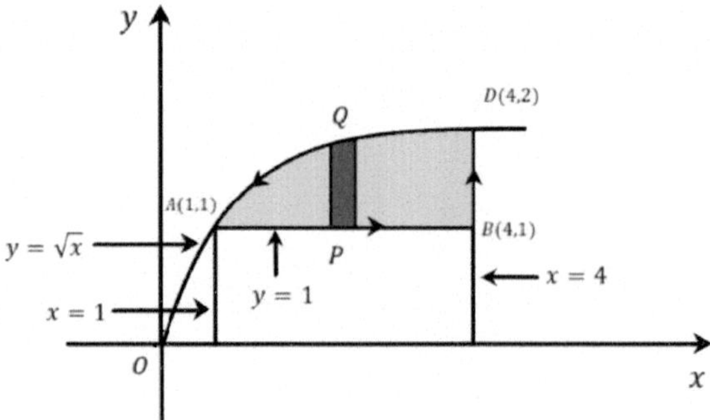

Figure 5.6 Represents the region bounded by the parabola and lines

$$= \int_1^4 dx = |x|_1^4 = 3$$

(b) Along $BD : x = 4 \Rightarrow dx = 0$ and y varies from 1 to 2.

$$\int_{BD} (Mdx + Ndy) = \int_{BD} \left(\frac{1}{y} dx + \frac{1}{x} dy \right)$$

$$= \int_1^2 \frac{1}{4} dy = \frac{1}{4} |y|_1^2 = \frac{1}{4}$$

(c) Along $DQA : y = \sqrt{x} \Rightarrow dy = \frac{1}{2\sqrt{x}} dx$ and x vary from 4 to 1.

$$\int_{DQA} (Mdx + Ndy) = \int_{DQA} \left(\frac{1}{y} dx + \frac{1}{x} dy \right)$$

$$= \int_4^1 \left(\frac{1}{\sqrt{x}} dx + \frac{1}{x} \cdot \frac{1}{2\sqrt{x}} dx \right)$$

$$= \left| 2\sqrt{x} - \frac{1}{\sqrt{x}} \right|_4^1$$

$$= 2 - 1 - 4 + \frac{1}{2} = -\frac{5}{2}$$

Substituting in equation (i), we get

$$\int_C (Mdx + Ndy) = 3 + \frac{1}{4} - \frac{5}{2} = \frac{3}{4} \qquad \text{(ii)}$$

(iv) Let R be the region bounded by the parabola $y = \sqrt{x}$ and the lines $x = 1, x = 4, y = 1$. Along the vertical strip, y varies from 1 to $\sqrt{x}$ and in the region R, x varies from 1 to 4.

$$\iint_R \left(\frac{\partial N}{\partial x} - \frac{\partial M}{\partial y} \right) dxdy = \int_1^4 \int_1^{\sqrt{x}} \left(-\frac{1}{x^2} + \frac{1}{y^2} \right) dxdy$$

$$= \int_1^4 \left| -\frac{1}{x^2} \cdot y - \frac{1}{y} \right|_1^{\sqrt{x}} dx$$

$$= \int_1^4 \left(-x^{-\frac{3}{2}} - x^{-\frac{1}{2}} + x^{-2} + 1 \right) dx$$

$$= \left| 2x^{-\frac{1}{2}} - 2x^{\frac{1}{2}} - \frac{1}{x} + x \right|_1^4$$

$$= 1 - 4 - \frac{1}{4} + 4 - 2 + 2 + 1 - 1$$

$$= \frac{3}{4} \tag{iii}$$

From equations (ii) and (iii), we see that

$$\oint_C Mdx + Ndy = \iint_R \left(\frac{\partial N}{\partial x} - \frac{\partial M}{\partial y} \right) dxdy$$

Hence, Green's theorem is verified.

5.2 Stokes' Theorem

Statement: If S be an open surface bounded by a closed curve C (See Figure 5.7) and $\overrightarrow{F}$ be a continuous and differentiable vector function, then

$$\oint_C \overrightarrow{F} \cdot d\overrightarrow{r} = \iint_S \nabla \times \overrightarrow{F} \cdot \hat{n}\, dS$$

where $\hat{n}$ is the unit outward normal at any point of the surface S.

Proof:

Let $\overrightarrow{F} = F_1\hat{i} + F_2\hat{j} + F_3\hat{k}$ and $\overrightarrow{r} = x\hat{i} + y\hat{j} + z\hat{k}$

$$\iint_S \nabla \times \overrightarrow{F} \cdot \hat{n}\, dS = \iint_S \nabla \times \left(F_1\hat{i} + F_2\hat{j} + F_3\hat{k} \right) \cdot \hat{n}\, dS$$

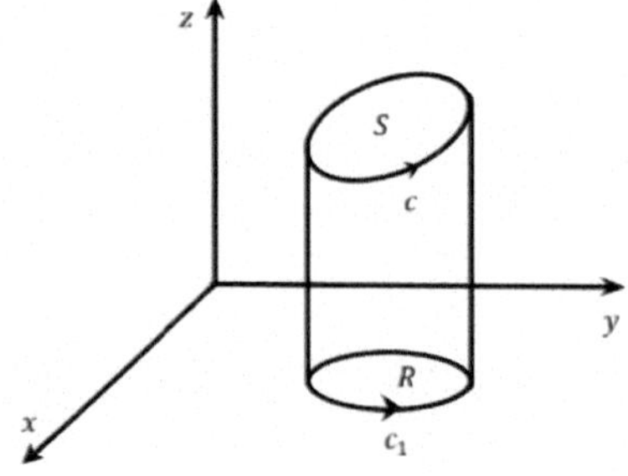

Figure 5.7 Represents an open surface bounded by a closed curve C

$$= \iint_S \left(\nabla \times F_1\hat{i}\right) \cdot \hat{n}\, dS + \iint_S \left(\nabla \times F_2\hat{j}\right) \cdot \hat{n}\, dS$$
$$+ \iint_S \left(\nabla \times F_3\hat{k}\right) \cdot \hat{n}\, dS \tag{5.3}$$

Consider,

$$\iint_S \left(\nabla \times F_1\hat{i}\right) \cdot \hat{n}\, dS = \iint_S \left[\left(\hat{i}\frac{\partial}{\partial x} + \hat{j}\frac{\partial}{\partial y} + \hat{k}\frac{\partial}{\partial z}\right) \times F_1\hat{i}\right] \cdot \hat{n}\, dS$$
$$= \iint_S \left(-\hat{k}\frac{\partial F_1}{\partial y} + \hat{j}\frac{\partial F_1}{\partial z}\right) \cdot \hat{n}\, dS$$
$$= \iint \left(\frac{\partial F_1}{\partial y}\hat{j} \cdot \hat{n} - \frac{\partial F_1}{\partial z}\hat{k} \cdot \hat{n}\right) dS \tag{5.4}$$

Let the equation of the surface S be $z = f(x, y)$,
Then,

$$\vec{r} = x\hat{i} + y\hat{j} + z\hat{k} \Rightarrow \vec{r} = x\hat{i} + y\hat{j} + f(x, y)\hat{k}$$

Differentiating partially with respect to y, we get

$$\frac{\partial \vec{r}}{\partial y} = \hat{j} + \frac{\partial f}{\partial y}\hat{k}$$

Taking dot product with $\hat{n}$,

$$\frac{\partial \vec{r}}{\partial y} \cdot \hat{n} = \hat{j} \cdot \hat{n} + \frac{\partial f}{\partial y}\hat{k} \cdot \hat{n} \tag{5.5}$$

$\frac{\partial \vec{r}}{\partial y}$ is tangential and $\hat{n}$ is normal to the surface S.

$$\frac{\partial \vec{r}}{\partial y} \cdot \hat{n} = 0$$

Substituting in the Equation (5.5),

$$0 = \hat{j} \cdot \hat{n} + \frac{\partial f}{\partial y}\hat{k} \cdot \hat{n}$$
$$\hat{j} \cdot \hat{n} = -\frac{\partial f}{\partial y}\hat{k} \cdot \hat{n} = -\frac{\partial z}{\partial y}\hat{k} \cdot \hat{n} \qquad (\because z = f(x, y))$$

Substituting in the Equation (5.4),

$$\iint_S \left(\nabla \times F_1\hat{i}\right) \cdot \hat{n}\, dS = \iint_S \left[\frac{\partial F_1}{\partial z}\left(-\frac{\partial z}{\partial y}\hat{j} \cdot \hat{n}\right) - \frac{\partial F_1}{\partial y}\hat{k} \cdot \hat{n}\right] dS$$

$$= -\iint_S \left(\frac{\partial F_1}{\partial z} \cdot \frac{\partial z}{\partial y} + \frac{\partial F_1}{\partial y}\right)\hat{k} \cdot \hat{n}\, dS \qquad (5.6)$$

The equation of the surface is $z = f(x, y)$.

$$F_1(x, y, z) = F_1[x, y, f(x, y)] = G(x, y)$$

Differentiating partially with respect to y,

$$\frac{\partial G}{\partial y} = \frac{\partial F_1}{\partial y} + \frac{\partial F_1}{\partial z} \cdot \frac{\partial z}{\partial y}$$

Substituting in the Equation (5.6),

$$\iint_S \left(\nabla \times F_1\hat{i}\right) \cdot \hat{n}\, dS = -\iint_S \frac{\partial G}{\partial y}\,\hat{k} \cdot \hat{n}\, dS$$

Let R is the projection of S on the xy-plane and $dxdy$ is the projection of dS on the xy-plane, then $\hat{k} \cdot \hat{n}\, dS = dxdy$
Thus,

$$\iint_S \left(\nabla \times F_1\hat{i}\right) \cdot \hat{n}\, dS = -\iint_R \frac{\partial G}{\partial y}\, dxdy = \oint_{C_1} Gdx$$

$$(\because \text{Using Green's theorem})$$

Since the value of G at each point (x, y) of C_1 is the same as the value of F_1 at each point (x, y, z) of C and dx is same for both the curves C_1 and C, we get

$$\iint_S \left(\nabla \times F_1\hat{i}\right) \cdot \hat{n}\, dS = \oint_C F_1 dx \qquad (5.7)$$

Similarly, by projecting the surface S on to yz and zx planes,

$$\iint_S \left(\nabla \times F_2\hat{j}\right) \cdot \hat{n}\, dS = \oint_C F_2 dy \qquad (5.8)$$

and

$$\iint_S \left(\nabla \times F_3\hat{k}\right) \cdot \hat{n}\, dS = \oint_C F_3 dz \qquad (5.9)$$

Substituting the Equations (5.7), (5.8), and (5.9) in (5.3),

$$\iint_S \left(\nabla \times \vec{F}\right) \cdot \hat{n}\, dS = \oint_C (F_1 dx + F_2 dy + F_3 dz) = \oint_C \left(\vec{F} \cdot d\vec{r}\right).$$

Note:
If surfaces S_1 and S_2 have the same bounding curve C, then

$$\iint_{S_1} \left(\nabla \times \vec{F}\right) \cdot \hat{n}\, dS = \iint_{S_2} \left(\nabla \times \vec{F}\right) \cdot \hat{n}\, dS = \oint_C \left(\vec{F} \cdot d\vec{r}\right).$$

Illustration 5.7: Verify Stokes's theorem for $\vec{F} = \left(x^2 + y^2\right)\hat{\imath} + 2xy\hat{\jmath}$, taken round the rectangle bounded by the lines $x = \pm a,\ y = 0,\ y = b$ (See Figure 5.8).

Solution:
Consider, $\vec{r} = x\hat{\imath} + y\hat{\jmath} \Rightarrow d\vec{r} = \hat{\imath}dx + \hat{\jmath}dy$ and let $ABCD$ be the given rectangle then

$$\int_{ABCD} \vec{F} \cdot d\vec{r} = \int_{AB} \vec{F} \cdot d\vec{r} + \int_{BC} \vec{F} \cdot d\vec{r} + \int_{CD} \vec{F} \cdot d\vec{r} + \int_{DA} \vec{F} \cdot d\vec{r}$$

and

$$\begin{aligned} \vec{F} \cdot d\vec{r} &= \left[\left(x^2 + y^2\right)\hat{\imath} + 2xy\hat{\jmath}\right] \cdot \left(\hat{\imath}dx + \hat{\jmath}dy\right) \\ &= \left(x^2 + y^2\right)dx + 2xydy \end{aligned}$$

Along $AB,\ x = a \Rightarrow dx = 0$ and y varies from 0 to b

$$\int_{AB} \vec{F} \cdot d\vec{r} = -2a \int_0^b ydy = -2a \cdot \frac{1}{2}b^2 = -ab^2$$

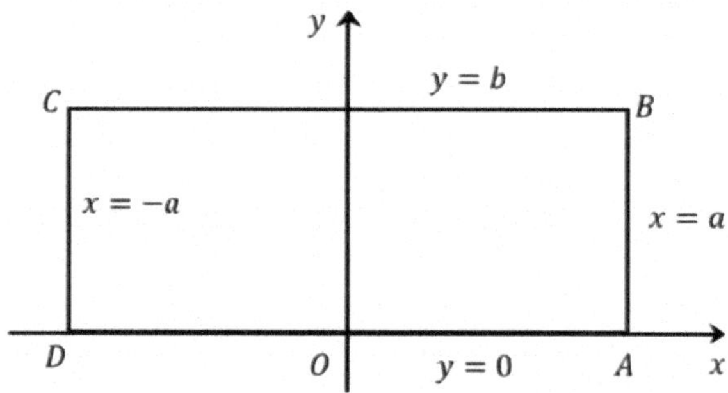

Figure 5.8 Represents the rectangle bounded by the lines

Similarly,

$$\int_{BC} \vec{F} \cdot d\vec{r} = \int_{-a}^{a} (x^2 + b^2)\, dx = -\frac{2}{3} \cdot a^2 - 2ab^2$$

$$\int_{CD} \vec{F} \cdot d\vec{r} = 2a \int_{b}^{0} ydy = -ab^2$$

and

$$\int_{DA} \vec{F} \cdot d\vec{r} = \int_{-a}^{a} x^2 dx = \frac{2a^3}{3}$$

Thus,

$$\int_{ABCD} \vec{F} \cdot d\vec{r} = -4ab^2.$$

Also, since curl $\cdot\, \vec{F} = -4y\hat{k}$,

$$\begin{aligned} \therefore \int_{S} \text{curl} \cdot \vec{F}\, dS &= \int_{0}^{b} \int_{-a}^{a} -4y\hat{k} \cdot \hat{k}\, dxdy \\ &= -4 \int_{0}^{b} \int_{-a}^{a} ydxdy \\ &= -4 \int_{0}^{b} |x|_{-a}^{a} ydy \\ &= -8a \left| \frac{y^2}{2} \right|_{0}^{b} = -4ab^2 \end{aligned}$$

Hence, the theorem is proved.

Illustration 5.8: Verify Stokes's theorem when $\vec{F} = (2x - y)\,\hat{\imath} - yz^2\hat{\jmath} - y^2z\hat{k}$, where S is the upper half surface of the sphere $x^2 + y^2 + z^2 = 1$ and C is the boundary.

Solution: Stokes's theorem is given as

$$\int_{C} \vec{F} \cdot d\vec{r} = \int_{S} \text{curl}\, \vec{F} \cdot \hat{n}\, dS$$

$$\therefore \quad \text{curl}\, \vec{F} = \begin{vmatrix} \hat{\imath} & \hat{\jmath} & \hat{k} \\ \partial/\partial x & \partial/\partial y & \partial/\partial z \\ 2x - y & -yz^2 & -y^2z \end{vmatrix} = \hat{k}$$

Let $\hat{n} = \sin\theta \cos\phi \;\hat{i} + \sin\theta \sin\phi \,\hat{j} + \cos\theta \,\hat{k}$

$$\therefore \quad \left(\text{curl } \vec{F}\right) \cdot \hat{n} = \cos\theta$$

$$\therefore \quad \int_S \text{curl } \vec{F} \cdot \hat{n} \, dS = \int_{\theta=0}^{\frac{\pi}{2}} \int_{\phi=0}^{2\pi} \cos\theta \sin\theta \;\, d\theta \, d\phi = \pi \qquad \text{(i)}$$

Again C is the unit circle $x^2 + y^2 = 1, \; z = 0$

$$\therefore x = \cos\phi \,, \quad y = \sin\phi \,, \quad z = 0$$

$$\therefore dx = -\sin\phi \;\, d\phi, \quad dy = \cos\phi \;\, d\phi, \quad dz = 0$$

$$\therefore \vec{F} \cdot d\vec{r} = \left[(2x - y)\,\hat{i} - yz^2\hat{j} - y^2 z\hat{k}\right] \cdot \left(\hat{i}dx + \hat{j}dy + \hat{k}dz\right)$$

$$= (2x - y)\,dx - yz^2 dy - y^2 z dz$$

$$= -\,(2\cos\phi \; - \sin\phi\,)\sin\phi \;\, d\phi$$

$$\therefore \quad \int_C \vec{F} \cdot d\vec{r} = -\int_0^{2\pi} (2\cos\phi \; - \sin\phi\,)\sin\phi \;\, d\phi$$

$$= -\int_0^{2\pi} \left(2\cos\phi \sin\phi \; - \sin^2\phi\,\right) \, d\phi$$

$$= \pi \qquad \text{(ii)}$$

From (i) and (ii), the theorem is verified.

Illustration 5.9: Verify Stokes's theorem for $\vec{F} \;=\; xy^2\hat{i} + y\hat{j} + z^2x\hat{k}$, for the surface of a rectangular lamina bounded by $x = 0, \; y = 0, \; x = 1, \; y = 2, \; z = 0.$

Solution: Here, we have $z = 0 \Rightarrow \vec{F} = xy^2\hat{i} + y\hat{j}$

$$\oint_C \vec{F} \cdot d\vec{r} = \oint_C \left[xy^2 dx + y dy\right]$$

where C is the path $OABCO$ as shown in Figure 5.9.
Along OA, $y = 0 \Rightarrow dy = 0$
Along AB, $x = 1 \Rightarrow dx = 0$
Along BC, $y = 2 \Rightarrow dy = 0$
Along CO, $x = 0 \Rightarrow dx = 0$

$$\therefore \quad \oint_C \vec{F} \cdot d\vec{r} = \int_{OA} xy^2 dx + \int_{AB} y dy + \int_{BC} xy^2 dx + \int_{CO} y dy$$

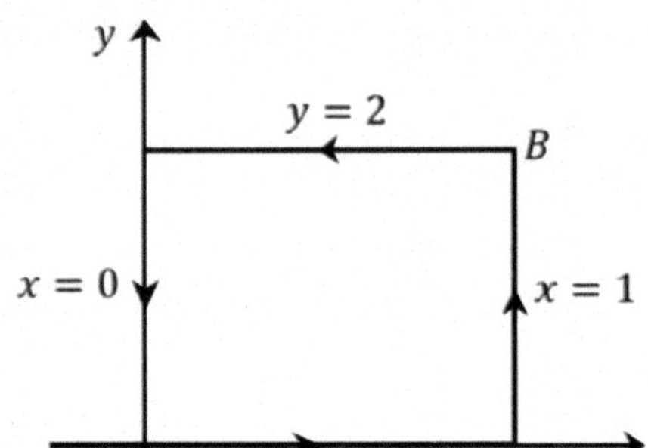

Figure 5.9 Represents the surface of a rectangular lamina bounded by the lines

$$= 0 + \int_0^2 y dy + \int_1^0 4x dx + \int_2^0 y dy$$

$$= \int_0^2 y dy + 4\left(\frac{x^2}{2}\right)_1^0 - \int_0^2 y dy \; = -2$$

Now,

$$\nabla \times \vec{F} = \begin{vmatrix} \hat{i} & \hat{j} & \hat{k} \\ \frac{\partial}{\partial x} & \frac{\partial}{\partial y} & \frac{\partial}{\partial z} \\ xy^2 & y & 0 \end{vmatrix} = \hat{i}\,(0) + \hat{j}\,(0) + \hat{k}\,(-2xy) = -2xy\hat{k}$$

Normal to the surface, $\hat{n} = \hat{k}$. $dS = dxdy$

$$\iint_S \left(\nabla \times \vec{F}\right) \cdot \hat{n}\; dS = \iint_S (-2xy)\,\hat{k} \cdot \hat{k} dxdy$$

$$= -2 \int_{x=0}^{1} \int_{y=0}^{2} xy\; dxdy$$

$$= -2 \int_0^1 x \left[\frac{y^2}{2}\right] dx$$

$$= -2 \int_0^1 x \left(\frac{4}{2}\right) dx$$

$$= -4 \left(\frac{x^2}{2}\right)_0^1 = -2$$

Thus,

$$\oint_C \vec{F} \cdot d\vec{r} = \iint_S \left(\nabla \times \vec{F}\right) \cdot \hat{n}\, dS = -2$$

which verifies Stokes's theorem.

Illustration 5.10: Use Stokes's theorem to evaluate $\int_C \vec{F} \cdot d\vec{r}$ if $\vec{F} = (x+y)\,\hat{i} + (2x-z)\,\hat{j} + (y+z)\,\hat{k}$ and C is the boundary of the triangle $(2,0,0)$, $(0,3,0)$ and $(0,0,6)$ (See Figure 5.10).

Solution:

$$\text{curl}\,\vec{F} = \begin{vmatrix} \hat{i} & \hat{j} & \hat{k} \\ \frac{\partial}{\partial x} & \frac{\partial}{\partial y} & \frac{\partial}{\partial z} \\ x+y & 2x-z & y+z \end{vmatrix} = 2\hat{i} + \hat{k}$$

Let $\phi = 3x + 2y + z = 6$

$$\therefore \hat{n} = \frac{\nabla\phi}{|\nabla\phi|} = \frac{3\hat{i} + 2\hat{j} + \hat{k}}{\sqrt{14}}$$

$$\text{curl}\,\vec{F} \cdot \hat{n} = \left(2\hat{i} + \hat{k}\right) \cdot \frac{3\hat{i} + 2\hat{j} + \hat{k}}{\sqrt{14}} = \frac{7}{\sqrt{14}}$$

Projection on xy-plane:

$$dS = \frac{dxdy}{\left|\hat{n} \cdot \hat{k}\right|} = \frac{\sqrt{14}}{1}\, dxdy$$

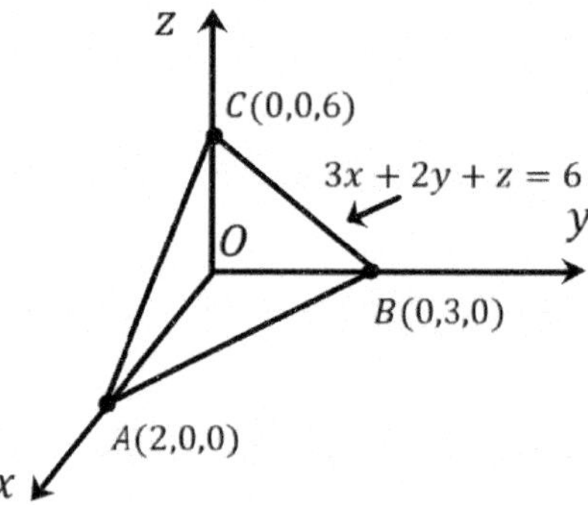

Figure 5.10 Represents the boundary of the triangle

$\therefore$ By Stokes's theorem we have,

$$\oint_C \vec{F} \cdot d\vec{r} = \iint_S \text{curl}\, \vec{F} \cdot \hat{n}\, dS = \iint_S \frac{7}{\sqrt{14}} \cdot \frac{\sqrt{14}}{1}\, dxdy$$
$$= \iint_S 7dxdy$$
$$= 7 \times \text{Area of the triangle on } xy\text{-plane}$$
$$= 7 \times \frac{1}{2} \times 3 \times 2 = 21$$

5.3 Gauss' Divergence Theorem

Statement: If $\vec{F}$ be a vector point function having continuous partial derivatives in the region bounded by a closed surface S, then

$$\iiint_V \nabla \cdot \vec{F}\, dV = \iint_S \vec{F} \cdot \hat{n}\, dS$$

where $\hat{n}$ is the unit outward normal at any point of the surface S.

Proof: Let $\vec{F} = F_1\hat{i} + F_2\hat{j} + F_3\hat{k}$,

$$\iiint_V \nabla \cdot \vec{F}\, dV = \iiint_V \left(\hat{i}\frac{\partial}{\partial x} + \hat{j}\frac{\partial}{\partial y} + \hat{k}\frac{\partial}{\partial z}\right)\left(F_1\hat{i} + F_2\hat{j} + F_3\hat{k}\right) dxdydz$$
$$= \iiint_V \left(\frac{\partial F_1}{\partial x} + \frac{\partial F_2}{\partial y} + \frac{\partial F_3}{\partial z}\right) dxdydz \qquad (5.10)$$

Assume a closed surface S such that any line parallel to the coordinate axes intersects S at most at two points. Divide the surface S into two parts: S_1- the lower and S_2- the upper part. Let $z = f_1(x, y)$ and $z = f_2(x, y)$ be the equations and $\hat{n}_1$ and $\hat{n}_2$ be normal to the surfaces S_1 and S_2 respectively (See Figure 5.11). Let R be the projection of the surface S on the xy-plane.

$$\iiint_V \frac{\partial F_3}{\partial z} dxdydz = \iint_R \left[\int_{f_1(x,y)}^{f_2(x,y)} \frac{\partial F_3}{\partial z} dz\right] dxdy$$
$$= \iint_R F_3(x, y, z)\,|_{f_1}^{f_2}\, dxdy$$
$$= \iint_R \left[F_3(x, y, f_2) - F_3(x, y, f_1)\right] dxdy$$

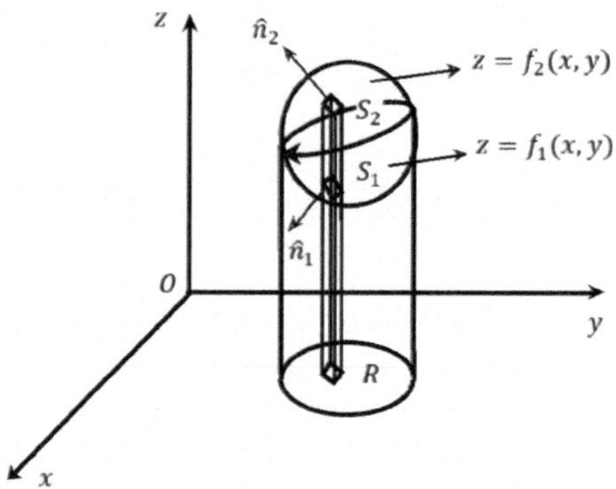

Figure 5.11 Represents the region bounded by a closed surface S

$$= \iint_R F_3\left(x, y, f_2\right)\, dxdy - \iint_R F_3\left(x, y, f_1\right)\, dxdy \tag{5.11}$$

Now, $dxdy =$ Projection of dS on xy-plane $= \hat{n} \cdot \hat{k}\, dS$
For surface $S_2 : z = f_2\left(x, y\right) \Rightarrow dxdy = \hat{n}_2 \cdot \hat{k}\, dS_2$
For surface $S_1 : z = f_1\left(x, y\right) \Rightarrow dxdy = -\hat{n}_1 \cdot \hat{k}\, dS_1$
Substituting in the Equation (5.11),

$$\begin{aligned} \iiint_V \frac{\partial F_3}{\partial z} dxdydz &= \iint_{S_2} F_3\, \hat{n}_2 \cdot \hat{k}\, dS_2 - \iint_{S_1} F_3\left(-\,\hat{n}_1 \cdot \hat{k}\right) dS_1 \\ &= \iint_{S_2} F_3\, \hat{n}_2 \cdot \hat{k}\, dS_2 + \iint_{S_1} F_3\;\hat{n}_1 \cdot \hat{k}\, dS_1 \\ &= \iint_S F_3\, \hat{n} \cdot \hat{k}\, dS \end{aligned} \tag{5.12}$$

Similarly, projecting the surface S on yz and $zx-$ planes, we get

$$\iiint_V \frac{\partial F_1}{\partial x} dxdydz = \iint_S F_1\, \hat{n} \cdot \hat{\imath}\, dS \tag{5.13}$$

and

$$\iiint_V \frac{\partial F_2}{\partial y} dxdydz = \iint_S F_2\, \hat{n} \cdot \hat{\jmath}\, dS \tag{5.14}$$

Substituting Equations (5.12), (5.13), and (5.14) in the Equation (5.10),

$$\iiint_V \nabla \cdot \vec{F}\, dV = \iint_S F_1\, \hat{n} \cdot \hat{\imath}\, dS + \iint_S F_2\, \hat{n} \cdot \hat{\jmath}\, dS + \iint_S F_3\, \hat{n} \cdot \hat{k}\, dS$$

$$= \iint_S \left(F_1\hat{i}\cdot\hat{n} + F_2\,\hat{j}\cdot\hat{n} + F_3\hat{k}\cdot\hat{n}\right)\,dS$$

$$= \iint_S \left(F_1\hat{i} + F_2\,\hat{j} + F_3\hat{k}\right)\cdot\hat{n}\,dS = \iint_S \vec{F}\cdot\hat{n}\,dS$$

Hence, $\iiint_V \nabla\cdot\vec{F}\,dV = \iint_S \vec{F}\cdot\hat{n}\,dS$

Note: The cartesian form of Gauss' divergence theorem is

$$\iiint_V \left(\frac{\partial F_1}{\partial x} + \frac{\partial F_2}{\partial y} + \frac{\partial F_3}{\partial z}\right) dxdydz = \iint_S (F_1dydz + F_2dzdx + F_3dxdy)$$

Illustration 5.11: Verify Gauss' divergence theorem for $\vec{F} = 4xz\hat{i} - y^2\hat{j} + yz\hat{k}$ over the cube $x = 0,\ x = 1,\ y = 0,\ y = 1,\ z = 0,\ z = 1.$

Solution: By Gauss's divergence theorem,

$$\iiint_V \nabla\cdot\vec{F}\,dV = \iint_S \vec{F}\cdot\hat{n}\,dS$$

(i) $\vec{F} = 4xz\hat{i} - y^2\hat{j} + yz\hat{k}$

$$\nabla\cdot\vec{F} = \frac{\partial}{\partial x}(4xz) + \frac{\partial}{\partial y}(-y^2) + \frac{\partial}{\partial z}(yz)$$

$$= 4z - 2y + y = 4z - y$$

(ii) For the cube (in Figure 5.12): x varies from 0 to 1

y varies from 0 to 1

z varies from 0 to 1

$$\iiint_V \nabla\cdot\vec{F}\,dV = \int_0^1\int_0^1\int_0^1 (4z - y)dxdydz$$

$$= \int_0^1\int_0^1 \left|2z^2 - yz\right|_0^1 dxdy$$

$$= \int_0^1 dx \int_0^1 (2 - y)dy$$

$$= |x|_0^1 \left|2y - \frac{y^2}{2}\right|_0^1$$

$$= 2 - \frac{1}{2} = \frac{3}{2} \qquad \text{(a)}$$

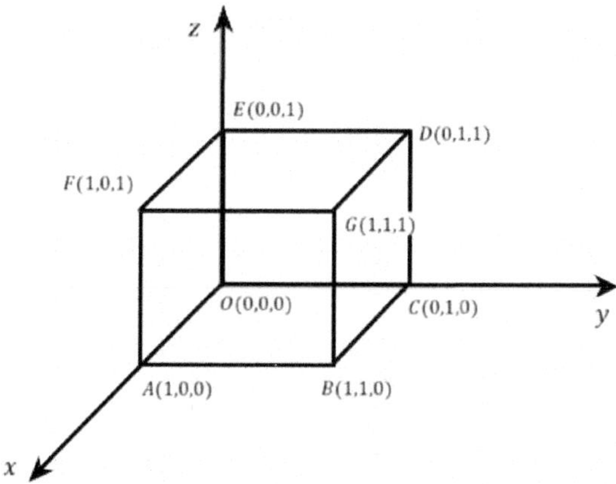

Figure 5.12 Represents the cube

(iii) Surface S of the cube consist of 6 surfaces, $S_1,\ S_2,\ S_3,\ S_4,\ S_4, S_5$, and S_6.

$$\therefore \iint_S \overrightarrow{F} \cdot \hat{n}\, dS = \iint_{S_1} \overrightarrow{F} \cdot \hat{n}\, dS + \iint_{S_2} \overrightarrow{F} \cdot \hat{n}\, dS + \iint_{S_3} \overrightarrow{F} \cdot \hat{n}\, dS + \iint_{S_4} \overrightarrow{F} \cdot \hat{n}\, dS + \iint_{S_5} \overrightarrow{F} \cdot \hat{n}\, dS + \iint_{S_6} \overrightarrow{F} \cdot \hat{n}\, dS \qquad \text{(b)}$$

1. On $S_1\,(OABC) : z = 0, \hat{n} = -\hat{k}, dS = \frac{dxdy}{|\hat{n}\cdot\hat{k}|} = dxdy$

   x and y both vary from 0 to 1.

$$\therefore \iint_{S_1} \overrightarrow{F} \cdot \hat{n}\, dS = \iint_{S_1} \left(4xz\hat{\imath} - y^2\hat{\jmath} + yz\hat{k}\right) \cdot \left(-\hat{k}\right) dxdy = 0$$

2. On $S_2\,(DEFG) : z = 1, \hat{n} = \hat{k}, dS = \frac{dxdy}{|\hat{n}\cdot\hat{k}|} = dxdy$

   x and y both vary from 0 to 1.

$$\therefore \iint_{S_2} \overrightarrow{F} \cdot \hat{n}\, dS = \iint_{S_2} \left(4xz\hat{\imath} - y^2\hat{\jmath} + yz\hat{k}\right) \cdot \left(\hat{k}\right) dxdy$$

$$= \int_0^1 \int_0^1 ydxdy = \int_0^1 \left|\frac{y^2}{2}\right|_0^1 = \frac{1}{2}|x|_0^1 = \frac{1}{2}$$

3. On $S_3\,(OAFE): y=0, \hat{n}=-\hat{j}, dS=\frac{dzdx}{|\hat{n}\cdot\hat{j}|}=dzdx$

   x and z both vary from 0 to 1.

$$\therefore \iint_{S_3} \vec{F}\cdot\hat{n}\,dS=\iint_{S_3}\left(4xz\hat{i}-y^2\hat{j}+yz\hat{k}\right)\cdot\left(-\hat{j}\right)dzdx=0$$

4. On $S_4\,(BCDG): y=1, \hat{n}=\hat{j}, dS=\frac{dzdx}{|\hat{n}\cdot\hat{j}|}=dzdx$

   x and z both vary from 0 to 1.

$$\therefore \iint_{s_4} \vec{F}\cdot\hat{n}\,dS=\iint_{S_4}\left(4xz\hat{i}-y^2\hat{j}+yz\hat{k}\right)\cdot\left(\hat{j}\right)dzdx$$

$$=\int_0^1\int_0^1 -dzdx=-1$$

5. On $S_5\,(OABC): x=0, \hat{n}=-\hat{i}, dS=\frac{dydz}{|\hat{n}\cdot\hat{i}|}=dydz$

   y and z both vary from 0 to 1.

$$\therefore \iint_{S_5} \vec{F}\cdot\hat{n}\,dS=\iint_{S_5}\left(4xz\hat{i}-y^2\hat{j}+yz\hat{k}\right)\cdot\left(-\hat{i}\right)dydz=0$$

6. On $S_6\,(ABGF): x=1, \hat{n}=\hat{i}, dS=\frac{dydz}{|\hat{n}\cdot\hat{i}|}=dydz$

   y and z both vary from 0 to 1.

$$\therefore \iint_{S_6} \vec{F}\cdot\hat{n}\,dS=\iint_{S_6}\left(4xz\hat{i}-y^2\hat{j}+yz\hat{k}\right)\cdot\left(\hat{i}\right)dydz$$

$$=\int_0^1\int_0^1 4zdzdx=2$$

Substituting in the equation (b),

$$\therefore \iint_S \vec{F}\cdot\hat{n}\,dS=0+\frac{1}{2}+0-1+0+2=\frac{3}{2} \qquad \text{(c)}$$

From equations (a) and (c), we get

$$\iiint_V \nabla\cdot\vec{F}\,dV=\iint_S \vec{F}\cdot\hat{n}\,dS=\frac{3}{2}$$

Hence, the Gauss theorem is verified.

Illustration 5.12: Verify Gauss' divergence theorem for $\overrightarrow{F} = 2xz\hat{i} + yz\hat{j} + z^2\hat{k}$ over the upper half of the sphere $x^2 + y^2 + z^2 = a^2$.

Solution:
By Gauss's divergence theorem,

$$\iiint_V \nabla \cdot \overrightarrow{F}\, dV = \iint_S \overrightarrow{F} \cdot \hat{n}\, dS$$

(i) $\overrightarrow{F} = 2xz\hat{i} + yz\hat{j} + z^2\hat{k}$

$$\nabla \cdot \overrightarrow{F} = \frac{\partial}{\partial x}(2xz) + \frac{\partial}{\partial y}(yz) + \frac{\partial}{\partial z}(z^2)$$

$$= 2z + z + 2z = 5z$$

(ii)

$$\iiint_V \nabla \cdot \overrightarrow{F}\, dV = \iiint_V 5z\, dxdydz$$

Taking $x = r\sin\theta\cos\phi$, $y = r\sin\theta\sin\phi$, $z = r\cos\theta$, equation of the sphere $x^2 + y^2 + z^2 = a^2$ reduces to $r = a$ and $dxdydz = r^2\sin\theta\ dr\, d\theta\, d\phi$.

For the upper half of the sphere (hemisphere in Figure 5.13),

r varies from 0 to a

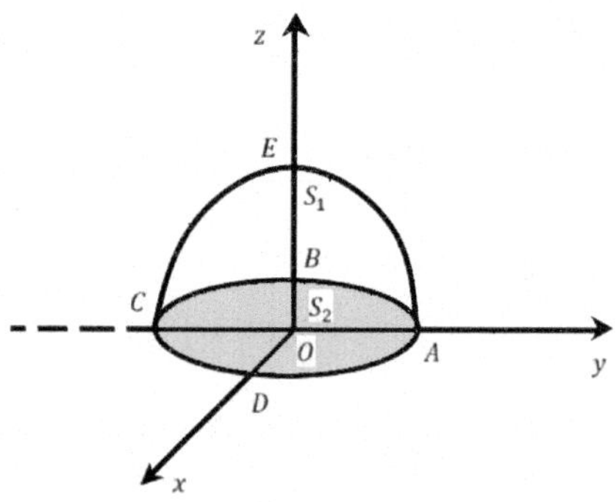

Figure 5.13 Represents the cube

θ varies from 0 to $\frac{\pi}{2}$

ϕ varies from 0 to 2π

$$\iiint_V \nabla \cdot \vec{F}\, dV = 5\int_{\phi=0}^{2\pi}\int_{\theta=0}^{\frac{\pi}{2}}\int_0^a r\cos\theta \cdot r^2\sin\theta\, dr d\theta d\phi$$

$$= 5\int_{\phi=0}^{2\pi} d\phi \int_{\theta=0}^{\frac{\pi}{2}} \cos\theta \sin\theta\ d\theta \left|\frac{r^4}{4}\right|_0^a$$

$$= 5|\phi|_0^{2\pi} \cdot \frac{1}{2}\left|-\frac{\cos 2\theta}{2}\right|_0^{\frac{\pi}{2}} \cdot \frac{a^4}{4} = \frac{5}{4}\pi a^4 \qquad \text{(a)}$$

(iii) The given surface is open and closed by the circular surface S_2in xy-plane. Thus, the surface S consists of two surfaces $S_1$and S_2.

$$\therefore \iint_S \vec{F} \cdot \hat{n}\, dS = \iint_{S_1} \vec{F} \cdot \hat{n}\, dS + \iint_{S_2} \vec{F} \cdot \hat{n}\, dS \qquad \text{(b)}$$

(1) Surface $S_1\,(ABCEA)$: This is the curved surface of the upper half of the sphere.
Let $\phi = x^2 + y^2 + z^2$

$$\therefore \hat{n} = \frac{\nabla\phi}{|\nabla\phi|} = \frac{2x\hat{i} + 2y\hat{j} + 2z\hat{k}}{\sqrt{4x^2 + 4y^2 + 4z^2}} = \frac{x\hat{i} + y\hat{j} + z\hat{k}}{a}$$

Let R be the projection of S_1 on the xy-plane, which is a circle $x^2 + y^2 = a^2$.

$$dS = \frac{dxdy}{\left|\hat{n} \cdot \hat{k}\right|} = \frac{adxdy}{z}$$

$$\iint_{S_1} \vec{F} \cdot \hat{n} dS$$

$$= \iint_R \left(2xz\hat{i} + yz\hat{j} + z^2\hat{k}\right) \cdot \left(\frac{x\hat{i} + y\hat{j} + z\hat{k}}{a}\right)\frac{adxdy}{z}$$

$$= \iint_R \left(2x^2 + y^2 + z^2\right) dxdy$$
$$= \iint_R \left(2x^2 + y^2 + a^2 - x^2 - y^2\right) dxdy$$
$$= \iint_R \left(x^2 + a^2\right) dxdy$$

Taking $x = r\cos\theta$, $y = r\sin\theta$,the equation of the circle $x^2+y^2 = a^2$ reduces to $r = a$ and $dxdy = r\ dr\ d\theta$. Along the radius vector OP, r varies from 0 to a and for the complete circle, θ varies from 0 to 2π (See Figure 5.14).

$$\iint_{S_1} \nabla \times \vec{F} \cdot \hat{n}\, dS = \int_0^{2\pi}\int_0^a \left(r^2\cos^2\theta\ + a^2\right) r\ dr\ d\theta$$
$$= \int_0^{2\pi} \left[\left|\frac{r^4}{4}\right|_0^a \cos^2\theta + a^2\left|\frac{r^2}{2}\right|_0^a\right] d\theta$$
$$= \int_0^{2\pi} \left[\frac{a^4}{4}\left(\frac{1+\cos2\theta}{2}\right) + \frac{a^4}{2}\right] d\theta$$
$$= a^4\left|\frac{5}{8}\theta + \frac{1}{8}\frac{\sin2\theta}{2}\right|_0^{2\pi} = \frac{5}{4}\pi a^4$$

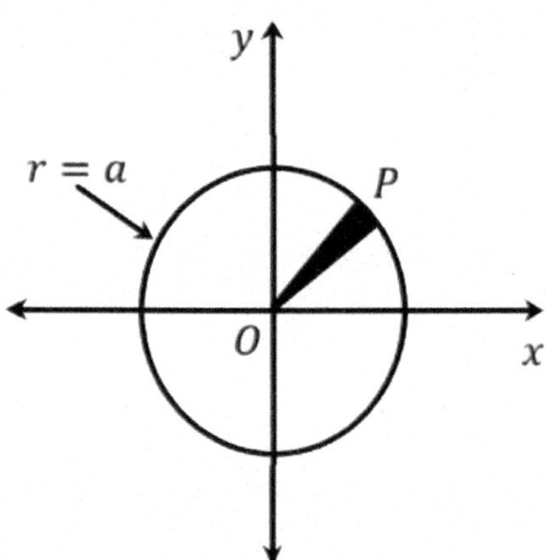

Figure 5.14 Represents the circle

(2) Surface $S_2\,(ABCDA)$: This is the circle $x^2 + y^2 = a^2$ in $xy-$plane $z = 0$, $\hat{n} = -\hat{k}$

$$\therefore dS = \frac{dxdy}{\left|\hat{n}\cdot\hat{k}\right|} = dxdy$$

$$\therefore \iint_{S_2} \vec{F}\cdot\hat{n}\,dS = \iint_{S_2} \left(2xz\hat{\imath} + yz\hat{\jmath} + z^2\hat{k}\right)\cdot\left(-\hat{k}\right) dxdy = 0$$

Substituting in Equation (b),

$$\therefore \iint_S \vec{F}\cdot\hat{n}\,dS = \frac{5}{4}\pi a^4 \tag{c}$$

From Equations (a) and (c), we have

$$\iiint_V \nabla\cdot\vec{F}dV = \iint_S \vec{F}\cdot\hat{n}\,dS = \frac{5}{4}\pi a^4$$

Hence, Gauss' theorem is verified.

Illustration 5.13: Evaluate $\iint_S \left(yz\,\hat{\imath} + zx\,\hat{\jmath} + xy\,\hat{k}\right)\cdot d\vec{S}$, where S is the surface of the sphere in the first octant.

Solution: By Gauss' divergence theorem,

$$\iint_S \vec{F}\cdot d\vec{S} = \iiint_V \nabla\cdot\vec{F}dV \tag{i}$$

$$\vec{F} = yz\,\hat{\imath} + zx\,\hat{\jmath} + xy\,\hat{k}$$

$$\therefore \nabla\cdot\vec{F} = \frac{\partial}{\partial x}(yz) + \frac{\partial}{\partial y}(zx) + \frac{\partial}{\partial z}(xy) = 0$$

From equation (i),

$$\iint_S \vec{F}\cdot d\vec{S} = \iiint_V \nabla\cdot\vec{F}dV = 0$$

Illustration 5.14: Evaluate $\iint_S (lx + my + nz)\;dS$, where l, m, n are the direction cosines of the outer normal to the surface whose radius is 2 units.

Solution: By Gauss' divergence theorem,

$$\iint_S \vec{F}\cdot\hat{n}\,dS = \iiint_V \nabla\cdot\vec{F}dV \tag{i}$$

(i) $\overrightarrow{F} \cdot \hat{n} = lx + my + nz$

$$= \left(x\,\hat{\imath} + y\,\hat{\jmath} + z\,\hat{k}\right) \cdot \left(l\,\hat{\imath} + m\,\hat{\jmath} + n\,\hat{k}\right)$$

$$\therefore \overrightarrow{F} = x\,\hat{\imath} + y\,\hat{\jmath} + z\,\hat{k}$$

(ii) $\therefore \nabla \cdot \overrightarrow{F} = \frac{\partial}{\partial x}(x) + \frac{\partial}{\partial y}(y) + \frac{\partial}{\partial z}(z) = 3$

(iii)

$$\iiint_V \nabla \cdot \overrightarrow{F} dV = \iiint_V 3\, dV$$

$$= 3\left(\begin{array}{c}\text{Volume of the region bounded by}\\ \text{the sphere of 2 unit radius}\end{array}\right)$$

$$= 3 \cdot \frac{4}{3}\,\pi(2)^3 = 32\,\pi$$

From Equation (i),

5.4 Exercise

1. Verify Green's theorem for the function $\overrightarrow{F} = \left(y^2 - 7y\right)\hat{\imath} + (2xy + 2x)\hat{\jmath}$ and curve $C : x^2 + y^2 = 1$.

   (Answer: 9π)

2. Verify Green's theorem for $\oint_C \left(3x - 8y^2\right) dx + (4y - 6xy)dy$ where C is the boundary of the triangle with vertices $(0,0)$, $(1,0)$ and $(0,1)$.

   $\left(\text{Answer} : \frac{5}{3}\right)$

3. Using Green's theorem, evaluate the linear integral $\oint_C(\sin y\ dx + \cos x dy)$ counterclockwise, where C is the boundary of the triangle with vertices $(0,0)$, $(\pi,\ 0)$, $(\pi, 1)$.

   $\left(\text{Answer} : -1 + \frac{2}{\pi} - \frac{1}{\pi}\cos 1\right)$

4. Evaluate $\oint_C \left[\left(x^2 - 2xy\right) dx + (x^2 y + 3)dy\right]$ around the boundary C of the region $y^2 = 8x,\ x = 2$.

   $\left(\text{Answer} : \frac{128}{5}\right)$

5. Verify Green's theorem in the plane for $\oint_C (2x - y^2)\, dx - xydy$ where C is the boundary of the region enclosed by the circles $x^2 + y^2 = 1$ and $x^2 + y^2 = 9$.

(Answer : 60π)

6. Verify Green's theorem for $\oint_C (x^2 - 2xy)\, dx + (x^2 y + 3)\, dy$ where C is the boundary of the region bounded by $y = x^2$ and the line $y = x$. (Answer : $\frac{1}{4}$)

7. Verify Stokes's theorem for $\overrightarrow{F} = (x^2 - y^2)\,\hat{\imath} + 2xy\hat{\jmath}$, in the rectangular region in the xy-plane bounded by the lines $x = \pm a,\ y = 0,\ y = b$.

(Answer : $4ab^2$)

8. Verify Stokes's theorem for $\overrightarrow{F} = (x + y)\,\hat{\imath} + (y + z)\,\hat{\jmath} - x\hat{k}$ and S is the surface of the plane $2x + y + z = 2$ which is in the first octant.

(Answer : -2)

9. Evaluate by Stokes's theorem $\oint_C (e^x dx + 2ydy - dz)$, where C is the curve $x^2 + y^2 = 4$, $z = 2$.

(Answer : 0)

10. Verify Stokes's theorem for $\overrightarrow{F} = xz\hat{\imath} + y\hat{\jmath} + y^2 x\hat{k}$ over the surface of the tetrahedron bounded by the planes $y = 0,\ z = 0$ and $4x + y + 2z = 4$ above the yz-plane.

(Answer : 0)

11. Verify Stokes's theorem for $\overrightarrow{F} = (x^2 + y + 2)\,\hat{\imath} + 2xy\hat{\jmath} + 4ze^x\hat{k}$ over the surface S of the paraboloid $z = 9 - (x^2 + y^2)$ above the xy-plane.

(Answer : -9π)

12. Evaluate using Stokes's theorem $\iint_S \nabla \times \overrightarrow{F} \cdot \hat{n}\, dS$ where $\overrightarrow{F} = 3y\hat{\imath} - xz\hat{\jmath} + yz^2\hat{k}$ and S is the surface of the paraboloid $x^2 + y^2 = 2z$ bounded by the plane $z = 2$ and C is its boundary traversed in the clockwise direction.

(Answer : 20π)

13. Prove that $\iint_S \frac{dS}{\sqrt{a^2x^2 + b^2y^2 + c^2z^2}} = \frac{4\pi}{\sqrt{abc}}$, where S is the ellipsoid $a^2x^2 + b^2y^2 + c^2z^2 = 1$.

14. Evaluate $\iint_S \vec{F} \cdot d\vec{S}$ using divergence theorem where $\vec{F} = x^3\,\hat{\imath} + y^3\,\hat{\jmath} + z^3\,\hat{k}$ and S is the surface of the sphere $x^2 + y^2 + z^2 = a^2$.

(Answer : $\frac{12}{5}\pi a^5$)

15. Evaluate $\iint \vec{F} \cdot d\vec{S}$ using Gauss' divergence theorem where $\vec{F} 2xy\,\hat{\imath} + yz^2\,\hat{\jmath} + zx\,\hat{k}$ and S is the surface of the region bounded by $x = 0$, $y = 0$, $z = 0$, $y = 3$, $x + 2z = 6$.

(Answer : $\frac{351}{2}$)

16. Verify Gauss' divergence theorem for $\vec{F} = (x^2 - yz)\hat{\imath} + (y^2 - zx)\hat{\jmath} + (z^2 - xy)\hat{k}$ over the region R bounded by the parallelepiped $0 \le x \le a$, $0 \le y \le b$, $0 \le z \le c$.

(Answer : $abc(a + b + c)$)

17. Verify Gauss' divergence theorem for $\vec{F} = x\hat{\imath} + y\hat{\jmath} + z\hat{k}$ over the region R bounded by the sphere $x^2 + y^2 + z^2 = 16$.

(Answer : 256π)

18. Verify Gauss' divergence theorem for $\vec{F} = 2xy\hat{\imath} + 6yz\hat{\jmath} + 3zx\hat{k}$ over the region bounded by the coordinate planes and the plane $x + y + z = 2$.

(Answer : $\frac{22}{3}$)

6

MATLAB Programming

6.1 Basic of MATLAB Programming

MATLAB is an acronym for MATrix LABoratory. MATLAB was created to make it simple to use the LINPACK (linear system package) and EISPACK (eigen system package) projects' matrix tools.

MATLAB, *see The MathWorks Inc., 2005, and D. Houcque, 2005,* is a high-performance technical computing language. It combines computing, visualization, and a programming environment into one package. MATLAB is also a contemporary programming language environment, with advanced data structures, built-in editing and debugging tools, and object-oriented programming capabilities. Because of these features, MATLAB is an outstanding teaching and research tool.

In this section, we presume that the reader is familiar with basic terminology like different windows in MATLAB, basic input commands, creating MATLAB variables, error messages, functions, etc.

6.1.1 Basic of MATLAB Programming

Many mathematical functions are predefined in MATLAB. By typing_help `elfun`, and help `specfun`, users can access inbuilt commands for elementary and special mathematical functions. A large number of mathematical functions can be evaluated using MATLAB. There are two types of arithmetic operations in MATLAB: array operations and matrix operations. Users may execute mathematical computations using these arithmetic operations, such as adding two integers, raising the elements of an array to a specific power, or multiplying two matrices.

6.1.1.1 Introductory MATLAB programmes

In this section, we discuss some basic examples of MATLAB programming to recall basic operations, functions, inputs, and mathematical computations. And it will help us in further MATLAB programs.

1. **Write a MATLAB program to find the addition, subtraction, and division of two numbers.**

   Solution: Let $a = 3$ and $b = 5$ be two numbers for these operations. Assume $c1$, $c2$, and $c3$ variables for storing these results. So, $c1$, $c2$, and $c3$ are the outputs for these operations respectively.

   Therefore

```
>> a=3;

>> b=5;

>> c1=a+b

c1 =

     8

>> c2=a-b

c2 =

    -2

>> c3=a/b

c3 =

    0.6000
```

   In this example, the semicolon at the end of the statements for a and b, suppress output to run the program smoothly. Whereas in the statement c1=a+b, c2=a-b, and c3=a/b are not followed by a semicolon so the stored values “dumped” as the output.

2. **Write a MATLAB program to evaluate the given expression for any two numbers.**

$$2a + b^2 - ab + \frac{b}{a} - 10$$

Solution: In this MATLAB program, write a user-defined program. Users define the programs are those programs in which the input of the programs are taken from the user. The main advantage of the user define programs are, the user can give any input rather than a fixed input. In this program, we use the "input" command to take values from the user. The syntax is given as follow:

```
variable = input('Enter input statement for
user');
```

Input:

```
a = input('Enter 1st number: \n');
b = input('Enter 2nd number: \n');
c = 2*a+b^2-a*b+b/a-10
```

Output:

```
Enter 1st number:
-10

Enter 2nd number:
9

c =

    140.1000
```

In this example, "\n" is given for the newline. i.e., once the program is run user can see the input statement, and the user can give input in the next line in the "command window".

3. **Write a MATLAB program to print "Welcome to MATLAB Programming"**

Solution:

Input:

>> fprintf ('Welcome to MATLAB Programming \n')

Output:

Welcome to MATLAB Programming

In this program, "fprintf (' ')" is used to print given input in a single quotation mark (' ').

4. **Write a user define MATLAB program to print square, cube, and square roots to preserve 4 digits after the decimal point for any number.**

Solution:

Input:

```
a = input('Enter the number : ');
b = a*a;
c = a*a*a;
d = sqrt(a);
fprintf(' The square of %4u is equals %4u. \r', a,
b)
fprintf(' The cube of %4u is equals %4u .\r', a,
c)
fprintf('The square root of %2u is %6.4f \r', a,
d)
```

Output:

1) If the number is 4 then

   Enter the number : 4

   The square of 4 equal 16.

   The cube of 4 equal 64.

   The square root of 4 is 2.0000.

2) If the number is 5 then

   Enter the number : 5

   The square of 5 equal 25.

   The cube of 5 equal 125.

   The square root of 5 is 2.2361.

In this program, "%4u" represents 4 digit integer value of the input number, and to preserve 4 digits after the decimal point we use "%6.4f".

5. **Write a MATLAB program to plot** $\sin x$ **for** $0 \leq x \leq 2\pi$.

Solution:

Input:

```
x=0:0.1:2*pi;
y=sin(x);
plot(x,y)
grid on
```

Output:

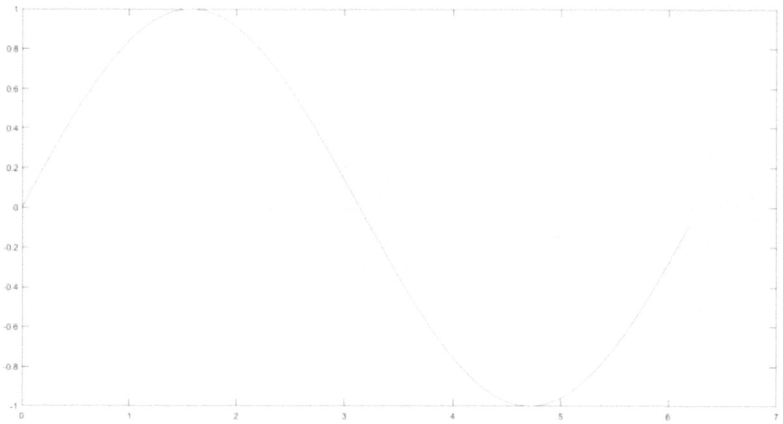

In this program, to plot $\sin x$ for $0 \leq x \leq 2\pi$, use the inbuilt command "plot(x,y)" then define x values using the command x=0:0.1:2*pi then store the values of $\sin x$ in variable y.

6. **Write a MATLAB program to draw any curve.**

Solution:

Input:

```
a = [0:0.01:5];
b = 4*a.^2 + 3*a -10;
plot(a,b)
grid on
```

Output:

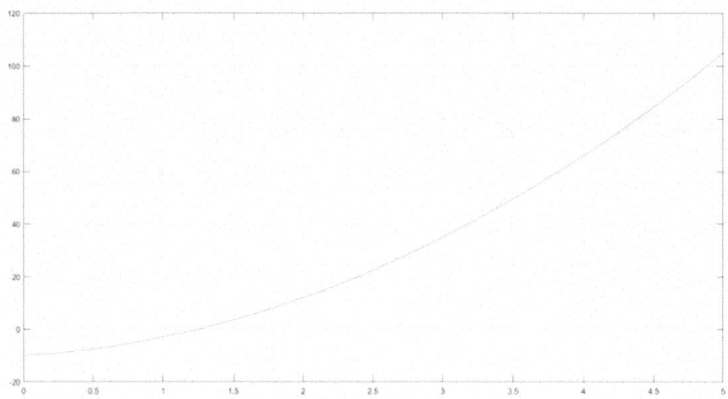

7. **Write a MATLAB program to plot user define line pattern, color, and thickness.**

   Solution:

   Input:

```
a = [0:0.5:5];
b = 2*a.^2 + 3*a -5;
plot(a,b,'-
ok','MarkerFaceColor','r','LineWidth',2)
xlabel('X'); ylabel('Y');
legend('Sample','Location','NorthWest')
grid on
```

   Output:

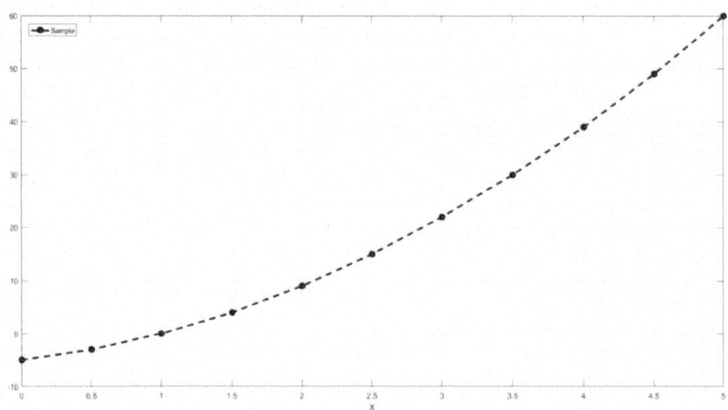

8. **Write a MATLAB program to plot multiple curves.**

Solution:

Input:

```
a = [0:0.5:5];
b = 2*a.^2 + 3*a -5;
c = 1.2*a.^2+4*a-3;
hold on
plot(a,b,'-
or','MarkerFaceColor','g','LineWidth',2)
plot(a,c,'--
ok','MarkerFaceColor','c','LineWidth',2)
xlabel('X'); ylabel('Y'); legend('Curve
1','Curve 2','Location','NorthWest')
grid on
```

Output:

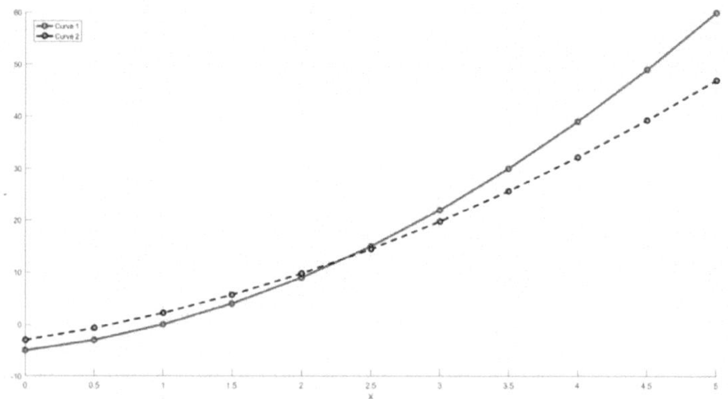

9. **Write a MATLAB program to plot multiple panels.**

Solution:

Input:

```
a = [0:0.5:5];
b = 2*a.^2 + 3*a -5;
c = 1.2*a.^2+4*a-3;
```

```
subplot(1,2,1)
plot(a,b,'-
or','MarkerFaceColor','g','LineWidth',2)
xlabel('X'); ylabel('Y'); legend('Curve
','Location','NorthWest')
grid on
subplot(1,2,2)
plot(a,c,'--
ok','MarkerFaceColor','c','LineWidth',2)
xlabel('X'); ylabel('Y'); legend('Curve
2','Location','NorthWest')
grid on
```

Output:

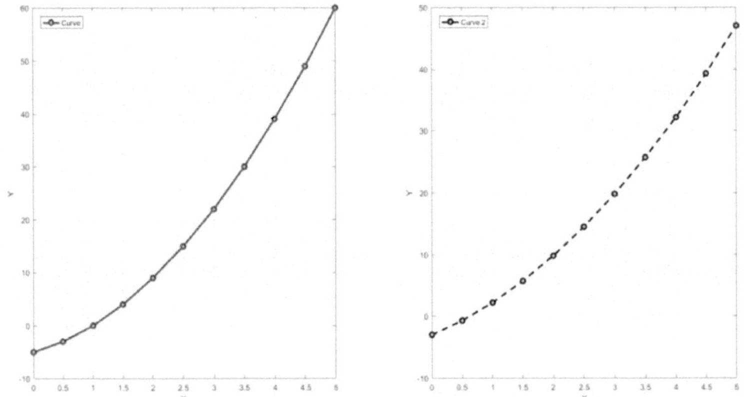

10. **Write a MATLAB program to plot a helix.**

Solution:

Input:

```
t = 0:0.01:5*pi;
plot3(sin(t),cos(t),t,'--rs','LineWidth',3,
'MarkerEdgeColor','k','MarkerFaceColor','g','Ma
rkerSize',1);
grid on
```

Output:

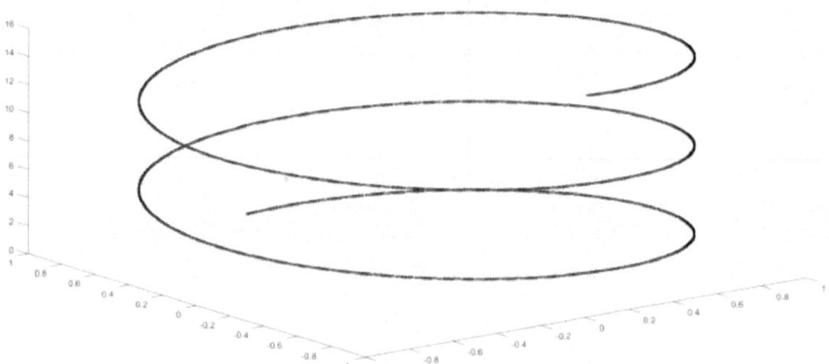

11. **Write a MATLAB program to print any number consecutive 5 times.**

Solution: In this program, we take input from the user as we have to print any number and use "for loop" for printing any number consecutive 5 times.

Input:

```
x = input('Enter the number : ');
for k = 1:5
x
end
```

Output:

Enter the number : 10

x =

    10

x =

    10

x =

    10

x =

```
        10
x =
        10
```

12. **Write a MATLAB program to print the utility of dummy index for any number consecutive 6 times. Also, take index from user.**

Solution:

Input:

```
b = input('Enter the number : ');
for k = 1:6
b^k
end
```

Output:

```
Enter the number : 2
ans =
        2
ans =
        4
ans =
        8
ans =
        16
ans =
        32
ans =
        64
```

Also, by user define index

Input

```
b = input('Enter the number : ');
x = input('Enter the index : ');
for k = 1:x
b^k
end
```

Output:

Enter the number : 2

Enter the index : 3

ans =

2

ans =

4

ans =

8

13. **Write a MATLAB program to find a sum upto 10.**

Solution:

Input:

```
sum1 = 0;
for k = 1:10
sum1 = sum1+k;
end
sum1
```

Output:

sum1 =

55

By user define number

Input:

```
sum1 = 0;
a=input('Enter a number upto which sum need to
find: ')
for k = 1:a
sum1 = sum1+k;
end
sum1
```

Output:

Enter a number upto which sum need to find: 5

a =

5

sum1 =

15

14. **Write a MATLAB program to find the sum of the first four elements in an array of six elements.**

Solution:

Input:

```
a = [3 -1 -1 4 10 5];
sum1 = 0;
for k = 1:4
sum1 = sum1+a(k);
end
sum1
```

Output:

sum1 =

5

By user define

```
a = input('Enter an array : ')
b = input('Enter a number to find sum upto
given number of elements : ')
sum1 = 0;
for k = 1:b
sum1 = sum1+a(k);
end
sum1
```

Output:

```
Enter an array : [1 2 3 4 5 6 7 8 9 10]
a =
   Columns 1 through 8
     1   2   3   4   5   6   7   8
   Columns 9 through 10
     9   10
 Enter a number to find sum upto the given number of elements: 7
b =
    7
sum1 =
   28
```

15. **Write a MATLAB a program for a double loop.**

Solution:

Input:

```
for n = 1:3
for m = 1:2
fprintf('n = %3u m = %3u \r', n, m)
end
end
```

Output:

```
n =   1 m =   1
n =   1 m =   2
n =   2 m =   1
n =   2 m =   2
n =   3 m =   1
n =   3 m =   2
```

16. **A MATLAB program for complicated use of loop and index.**

Solution:

Input:

```
a = [2 5 7 4 9 8 3];
b = [2 3 5 7];
sum1 = 0;
for k = 1:4
sum1 = sum1+a(b(k));
end
sum1
```

Output:

sum1 =

24

Summation performed by this program

sum1 = a(b(1))+a(b(2))+a(b(3))+a(b(4))

= a(2)+a(3)+a(5)+a(7)

= 5+7+9+3

= 24

17. **Write a MATLAB program to find greater than or less than or equals to of any two numbers.**

Solution:

Input:

```
a = input('Enter 1st number : ')
b = input('Enter 2nd number : ')
if (a > b)
fprintf('%4u is greater than %4u \r', a,b)
else
fprintf('%4u is less than or equal to %4u \r',
a,b)
end
```

Output 1:

Enter 1st number : 10

a =

10

Enter 2nd number : 12

b =
12

10 is less than or equal to 12

Output 2:

Enter 1st number : 10

a =

10

Enter 2nd number : -1

b =

-1

10 is greater than -1.000000e+00

18. **Write a MATLAB program to represent the use of "if - elseif - else" loop.**

Solution:

Input:

```
a = 4;
if (a >= 5)
fprintf('%4i is greater than or equal to 5 \r',
a)
elseif (a > 1)
```

```
fprintf('%4i is less than 5 but greater than 1
\r', a)
elseif (a == 1)
fprintf('%4i equals 1 \r', a)
elseif (a > -3)
fprintf('%4i is less than 1 but greater than -3
\r', a)
else
fprintf('%4i is less than or equal to -3 \r',
a)
end
```

Output:

4 is less than 5 but greater than 1

19. **Write a MATLAB program to determine whether a given year is a leap year or not.**

Solution:

Input:

```
y = input('Enter a year in YYYY format : ');
if (mod(y, 400) == 0)
fprintf('%d is a leap year', y)
elseif (mod(y,4) == 0) & (mod(y,100) ~= 0)
fprintf('%d is a leap year', y)
else
fprintf('%d is not a leap year', y)
end
```

Output:

Enter a year in YYYY format : 2020

2020 is a leap year

Enter a year in YYYY format : 2022

2022 is not a leap year

6.1.1.2 Representation of a Vector in MATLAB

In this section, we represent a vector as a special case of a matrix. An array of dimension $1 \times n$ is called a row vector and $m \times 1$ dimensional array is called a column vector. The elements of a vector are enclosed by square brackets. The elements of a row vector are separated using spaces or by commas, whereas the elements of a column vector are separated using a semicolon(;).

For example, for a row vector v1 and v2

```
>> v1=[1 2 3 4 5]

v1 =

     1   2   3   4   5

>> v2=[1, 2, 3, 4, 5]

v2 =

     1   2   3   4   5
```

And for a column vector cv1 and cv2

```
>> cv1=[1;-4;3;9;-10]

cv1 =

     1
    -4
     3
     9
   -10

>> cv2=[1;2;-2;0.1]
```

```
cv2 =

    1.0000
    2.0000
   -2.0000
    0.1000
```

The transcribe of a row or a column vector is obtained using an apostrophe or a single quote (').

For example, transpose of a row vector v1 and a column vector cv1

```
>> w=v1'

w =

     1
     2
     3
     4
     5

>> w1=cv1'

w1 =
     1    -4     3     9   -10
```

To access elements, we use colon notation (:). For example, for the first three elements of the w1 vector,

```
>> w1(1:3)

ans =

     1    -4     3
```

Or, all elements after 3rd element

```
>> w1(3:end)


ans =

    3    9  -10
```

Similarly, we can find elements in the column vector also.

20 **Write a MATLAB program to find the addition of two vectors.**

Solution:

Input:

```
a = [2 12 25];
b = [3 7 4];
c = a+b
```

Output:

```
c =
  5    19    29
```

21 **Write a MATLAB program to find the multiplication of a vector by a scalar.**

Solution:

Input:

```
a = [2 12 25];
b = 4
c = a*b
```

Output:

```
c =
      8    48   100
```

6.1.1.3 Representation of a Matrix in MATLAB

A matrix is an array of numbers. We use the following steps to enter a matrix in MATLAB.

1. We begin with a left square bracket. i.e., [
2. Separate each element in a row with space or comma. i.e., (,)
3. Use a semicolon (;) to separate each row in the matrix.
4. After adding all rows in a matrix, end matrix with the right square bracket. i.e.,]

For example, to enter a matrix

$$A = \begin{bmatrix} 1 & 2 \\ 3 & 4 \end{bmatrix}$$

We use

```
>> A=[1 2; 3 4]
```

and the output matrix in MATLAB will be

```
A =
    1  2
    3  4
```

For non-square matrix

$$A = \begin{bmatrix} 1 & 2 \\ 3 & 4 \\ 5 & 6 \end{bmatrix}$$

We have,

```
>> A=[1, 2; 3, 4; 5, 6]

A =
    1  2
    3  4
    5  6
```

Once the matrix is stored in MATLAB, we can view a particular element of the matrix. For example, to view elements in row 1st and column 2nd, we use A(2,1)

```
>> A(2,1)

ans =

     3
```

To find the dimension of a vector or a matrix, we use the size command. For example, to find the dimension of the matrix

$$A = \begin{bmatrix} 1 & 2 \\ 3 & 4 \\ 5 & 6 \end{bmatrix}$$

We have

```
>> size(A)

ans =

     3     2
```

For a square matrix

$$A = \begin{bmatrix} 1 & 2 \\ 3 & 4 \end{bmatrix}$$

We have

```
>> size(A)

ans =

     2     2
```

i.e., size(A)=[m n]

To find the transpose of a matrix, we can use an apostrophe or a single quote ('). For example, transpose of a matrix

$$B = \begin{bmatrix} 0 & 1 & 3 \\ -2 & 0 & 7 \end{bmatrix}$$

We have

B =

0 1 3

-2 0 7

>> B'

ans =
0 -2
1 0
3 7

6.2 Some Miscellaneous Examples using MATLAB Programming

1. `% Plot the vectors (1,4,1), (1,-2,1) and (1,2,-6) directed from the origin.`

Solution:

```
point1 = [1,4,1];
point2 = [1,-2,1];
point3 = [1 2 -6];
origin = [0,0,0];
figure;hold on;
plot3([origin(1) point1(1)],[origin(2)
point1(2)],[origin(3) point1(3)],'r-^',
'LineWidth',2);
plot3([origin(1) point2(1)],[origin(2)
point2(2)],[origin(3) point2(3)],'g-^',
'LineWidth',2);
plot3([origin(1) point3(1)],[origin(2)
point3(2)],[origin(3) point3(3)],'b-^',
'LineWidth',2);
grid on;
xlabel('X axis'), ylabel('Y axis'), zlabel('Z
axis')
set(gca,'CameraPosition',[1 2 3]);
```

Output:

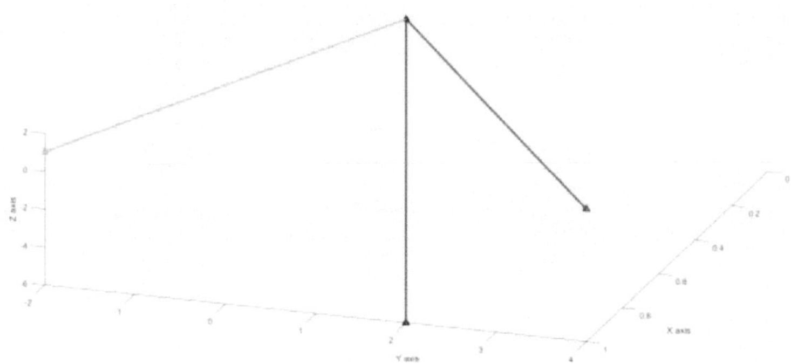

2. `%Find the modulus of the vector (3,4,-5).`

Solution:

```
x=input('x-coordinate of a vector: ');
y=input('y-coordinate of a vector: ');
z=input('z-coordinate of a vector: ');
m1=sqrt(x*x+y*y+z*z);

fprintf('The modulus of the vector (%d,%d,%d) =
%f',x,y,z,m1);
```

Input:

x-coordinate of a vector: 3
y-coordinate of a vector: 4
z-coordinate of a vector: -5

Output:

The modulus of the vector (3,4,-5) = 7.071068

3. `%If a=3i-j-4k, b=-2i+4j-3k and c=i+2j-k, then find the vector 3a-2b+4c.`

Solution:

```
a = [3 -1 -4];
b = [-2 4 -3];
c = [1  2 -1];
```

```
%then the vector 3a-2b+4c is
z=3*a-2*b+4*c
```

Output:

```
z =
  17     -3     -10
```

4. %If a=(1,2,1), b=(2,1,1) and c=(3,4,1) then find |a+2b+c|.

Solution:

```
a = [1  2  1];
b = [2  1  1];
c = [3  4  1];

%then |a+2b+c|

z1 = a+2*b+c

z2 = z1.*z1

z3 = sum(z2)

z4 = sqrt(z3)
```

Output:

```
z1 =
      8    8    4

z2 =

     64   64   16

z3 =

     144

z4 =

      12
```

5. % Identify whether the given vector is a unit or null vector or not.

Solution:

```
x=input('x-coordinate of a vector: ');
y=input('y-coordinate of a vector: ');
z=input('z-coordinate of a vector: ');

m1=sqrt(x*x+y*y+z*z);

if m1==1

fprintf('The vector (%d, %d, %d) is a unit vector.
\n',x,y,z);

elseif m1==0
    fprintf('The vector (%d, %d, %d) is a zero
vector. \n',x,y,z);

else
    fprintf('The vector (%d, %d, %d) is a not a
unit vector. \n',x,y,z);
    end
```

Input-1:

x-coordinate of a vector: 1
y-coordinate of a vector: 0
z-coordinate of a vector: 0

Output-1:

The vector (1, 0, 0) is a unit vector.

Input-2:

x-coordinate of a vector: 1
y-coordinate of a vector: 1
z-coordinate of a vector: 0

Output-2:

The vector (1, 1, 0) is not a unit vector.

Input-3:

x-coordinate of a vector: 0
y-coordinate of a vector: 0
z-coordinate of a vector: 0

Output-3:
The vector (0, 0, 0) is a zero vector.

Input-4:
x-coordinate of a vector: 1/sqrt(3)
y-coordinate of a vector: 0
z-coordinate of a vector: sqrt(2/3)

Output-4:
The vector (5.773503e-01, 0, 8.164966e-01) is a unit vector.

6. `%If v1=(-1,6,8) and v2=(1,3,-4) then find addition and subtraction.`

Solution:

```
v1=input('The first vector in the form of [x y z]: ');
v2=input('The Second vector in the form of [x y z]: ');

%Suppose z1 is a new addition vector then
%Addition
z1=v1+v2

%Suppose z2 is a new Subtractive vector then
%Subtraction
z2=v1-v2
```

Input:
The first vector in the form of [x y z]: [-1 6 8]
The second vector in the form of [x y z]: [1 3 -4]

Output:

```
z1 =

    0    9    4

z2 =

   -2    3   12
```

7. %If an arbitrary vector is given then find a product of an arbitrary vector with an arbitrary scalar.

Solution:

```
%Let a=(x,y,z) be an arbitrary vector then
x=input('Enter x-coordinate of an arbitrary vector
: ');
y=input('Enter y-coordinate of an arbitrary vector
: ');
z=input('Enter z-coordinate of an arbitrary vector
: ');
d=input('Enter  an  arbitrary  scalar  for  product
with given arbitrary vector : ');

a=[x y z]

da=d*a
```

Input:

Enter x-coordinate of an arbitrary vector : 3
Enter y-coordinate of an arbitrary vector : 4
Enter z-coordinate of an arbitrary vector : 5
Enter an arbitrary scalar for a product with a given arbitrary vector : -1

Output:

a =

3 4 5

da =

-3 -4 -5

```
8. %If x=(1,2,3) and y=(2,3,4),then find (i) x.y
and (ii) Angle between x and y in degree.
```

Solution:

```
x=[1 2 3];
y=[2 3 4];

x1=sqrt(sum(x.*x));
y1=sqrt(sum(y.*y));

z1=sum(x.*y)

z2=acosd(z1/x1*y1)

fprintf('x.y =%d \n',z1);
fprintf('The angle between x and y is %f .\n',z2);
```

Output:

z1 =

20

z2 =

0.0000e+00 + 2.3220e+02i

x.y = 20

The angle between x and y is 0.000000.

```
8. % Write a user input programme to find (i)
   x.y and (ii) Angle between x and y in
   degree.
```

Solution:

```
x=input('Enter a first vector in the format [x y
z] = ');
y=input('Enter a second vector in the format [x y
z] = ');
```

```
x1=sqrt(sum(x.*x));
y1=sqrt(sum(y.*y));

z1=sum(x.*y)

z2=acosd(z1/x1*y1)

fprintf('x.y =%d \n',z1);
fprintf('The angle between x and y is %f \n',z2);
```

Input 1:

Enter a first vector in the format [x y z] = [1 2 3]
Enter a second vector in the format [x y z] = [-1 -2 -3]

Output 1:

z1 =

-14

z2 =

1.8000e+02 - 1.9085e+02i

x.y =-14

The angle between x and y is 180.000000

Input 2:
Enter a first vector in the format [x y z] = [1 0 0]
Enter a second vector in the format [x y z] = [0 1 0]

Output 2:

z1 =

0

z2 =
90

x.y =0

The angle between x and y is 90.000000

Input 3:
Enter a first vector in the format [x y z] = [1 -5 6]
Enter a second vector in the format [x y z] = [6 3 -1]

Output 3:

z1 =

-15

z2 =

1.8000e+02 - 1.8624e+02i

x.y =-15

The angle between x and y is 180.000000

```
9. %If x=(1,-2,2) and y=(0,0,-1), then verify
   that
%(i)|x.y|<=|x||y|
%(ii)|x+y|<=|x|+|y|
%(iii)|x-y|>=|x|-|y|
```

Solution:

```
x=input('Enter a first vector in the format [x y
z] = ');
y=input('Enter a second vector in the format [x y
z] = ');

x1=sqrt(sum(x.*x));
y1=sqrt(sum(y.*y));
e1=sqrt(sum((x.*y).*(x.*y)));
e2=sqrt(sum((x+y).*(x+y)));
e3=x1.*y1;
e4=x1+y1;
e5=sqrt(sum((x-y).*(x-y)));
e6=x1-y1;

if e1<=e3
    fprintf('(i) is true. %f <= %f \n', e1, e3);
```

```
else
    fprintf('(i) is false.');
end
if e2<=e4
    fprintf('(ii) is true. %f <= %f \n', e2, e4);
else
    fprintf('(ii) is false.');
end
if e5>=e6
    fprintf('(iii) is true. %f >= %f \n', e5, e6);
else
    fprintf('(iii) is false.');
end
```

Input 1:

Enter a first vector in the format [x y z] = [1 -2 2]
Enter a second vector in the format [x y z] = [0 0 -1]

Output 1:
(i) is true. 2.000000 < = 3.000000
(ii) is true. 2.449490 < = 4.000000
(iii) is true. 3.741657 > = 2.000000

Input 2:
Enter a first vector in the format [x y z] = [1 -1 -1]
Enter a second vector in the format [x y z] = [2 -4 7]

Output 2:

(i) is true. 8.306624 < = 14.387495

(ii) is true. 8.366600 < = 10.038675

(iii) is true. 8.602325 > = -6.574573

```
    10.%Write a user define programme to find
%(i) The projection of a on b.
%(ii)The projection of b on a.
```

Solution:

```
a=input('Enter a first vector in the format [x y
z] = \n');
b=input('Enter a second vector in the format [x y
z] = \n');

x1=sqrt(sum(a.*a));
```

```
y1=sqrt(sum(b.*b));
e1=sum(a.*b);

p1=e1/y1;
p2=e1/x1;

fprintf('The projection of a on b is %f .\n',p1);
fprintf('The projection of b on a is %f .\n',p2);
```

Input:

Enter a first vector in the format [x y z] =
[3 -2 1]
Enter a second vector in the format [x y z] =
[1 -2 1]

Output:

The projection of a on b is 3.265986.
The projection of b on a is 2.138090.

11. % Write a user define programme to verify whether %two vectors are
% perpendicular or not.

Solution:

```
a=input('Enter a first vector in the format [x y
z] = \n');
b=input('Enter a second vector in the format [x y
z] = \n');

c=dot(a,b)

if c==0

fprintf('The vectors a and b are perpendicular to
each other.');
else
fprintf('The vectors a and b are not perpendicular
to each other.');
end
```

Input 1:

Enter a first vector in the format [x y z] =
[1 0 0]
Enter a second vector in the format [x y z] =

[0 1 0]

Output 1:

c =

0

The vectors a and b are perpendicular to each other.

Input 2:

Enter a first vector in the format [x y z] =
[1 0 0]
Enter a second vector in the format [x y z] =
[1 0 0]

Output 2:

c =

1

The vectors a and b are not perpendicular to each other.

Input 3:

Enter a first vector in the format [x y z] =
[1 2 3]
Enter a second vector in the format [x y z] =
[-2 1 3]

Output 3:

c =

9

The vectors a and b are not perpendicular to each other.

12. % Write a user define programme to find the vector or cross product of two vectors

Solution:

```
a=input('Enter a first vector in the format [x y
z] = \n');
b=input('Enter a second vector in the format [x y
z] = \n');

c=cross(a,b)

fprintf('The cross product is (%d,%d,%d).',
c(1,1), c(1,2),c(1,3));
```

Input:

Enter a first vector in the format [x y z] =
[2 -3 1]
Enter a second vector in the format [x y z] =
[1 -1 2]

Output:

```
c =
      -5    -3     1
```

The cross product is (-5, -3, 1).

```
13. % Write a program for forces 3i-j+2k
      and i+3j-k act on a particle and
% the particle moves from 2i+3j+k to 5i+2j+k under
these forces.
% Find the work done by these forces.
```

Solution:

```
f1=[3 -1 2];
f2=[1 3 -1];
a=[2 3 1];
b=[5 2 1];
d=b-a;

f=f1+f2;

WorkDone=dot(f,d);

fprintf('The total work done is %f
units.\n',WorkDone);
```

Output:

The total work done is 10.000000 units.

```
14. % Find the moment about the point i- 3j+k
      of the force 4i-3k
% acting through the point (2,-2,5).
% Also, find the magnitude of the moment.
```

Solution:

```
A=[1 -3 1];
B=[2 -2 5];
F=[4 0 -3];
```

```
D=B-A;

M=cross(D,F);

fprintf('The moment of the force F about the point
A is (%d, %d, %d) .\n',M(1,1),M(1,2),M(1,3));
M1=sqrt(sum(M.*M));

fprintf('The magnitude of the moment is %f
.\n',M1);
```

Output:

The moment of the force F about the point A is (-3, 19, -4).
The magnitude of the moment is 19.646883.

```
15. % Write a MATLAB programe to find arc
      length of the curve and
% x(t)=sin(2t),y(t)=cos(t),z(t)=t, where t lies in
[0,3pi].
% also plot the curve
```

Solution:

```
t = 0:0.1:3*pi;
plot3(sin(2*t),cos(t),t)
f = @(t) sqrt(4*cos(2*t).^2 + sin(t).^2 + 1);
len = integral(f,0,3*pi)
```

Output:

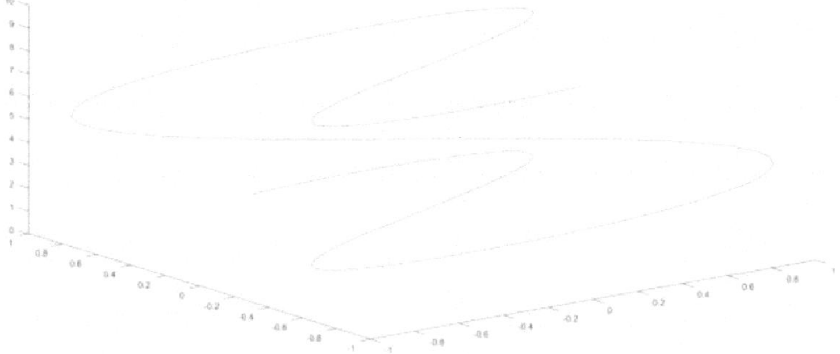

len =

17.2220

```
16. % Plot the 2D gradient for the
      following function.
```

$$xe^{-x^2-y^2}$$

Solution:

```
v = -2:0.1:2;
[x,y] = meshgrid(v);
z = x .* exp(-x.^2 - y.^2);
[px,py] = gradient(z,.5,.5);
contour(v,v,z)
hold on
quiver(v,v,px,py)
hold off
```

Output:

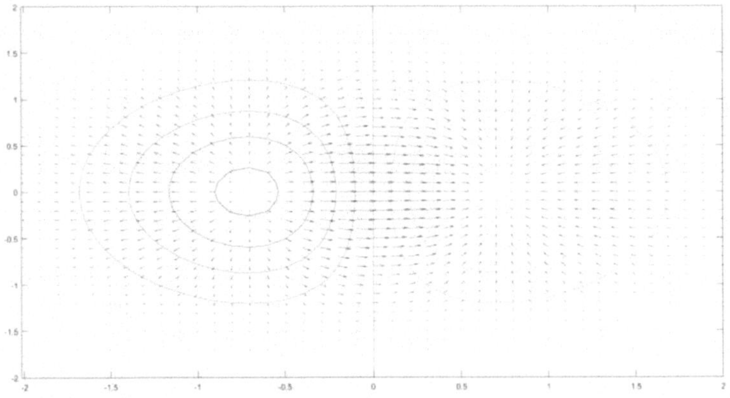

For the function:

$3x^2y - y^3z^2$

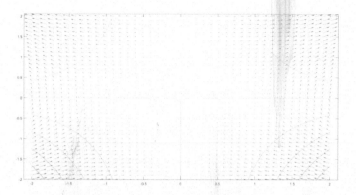

17. %Find the line integral for the following function

$$-t\sin t + \cos^2 t + \sin t$$

From o to 2π.

Solution:

t = 0:0.1:2*pi;
f = @(t)(-t.*sin(t)+cos(t).∧2+sin(t));
len = integral(f,0,2*pi)

Output:

len =

9.4248

18. %Programe for the curl angular velocity in one plane of the volume
%plots the velocity vectors in the same plane.

Solution:

```
load wind
k = 4;
x = x(:,:,k); y = y(:,:,k); u = u(:,:,k); v =
v(:,:,k);
cav = curl(x,y,u,v);
pcolor(x,y,cav); shading interp
hold on;
quiver(x,y,u,v,'b')
hold off
colormap hot(60)
```

Output:

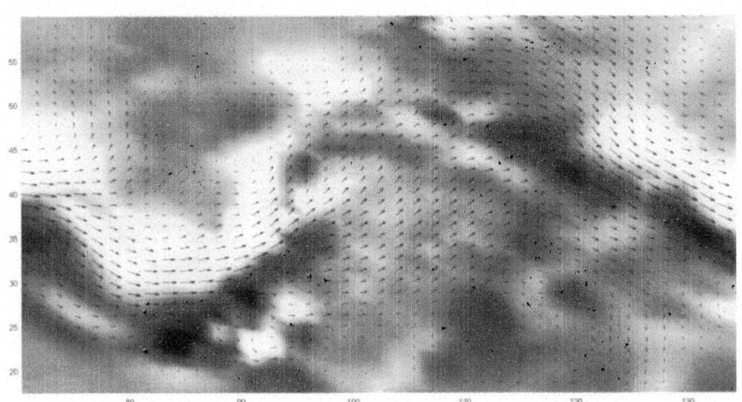

19. Find volume integral for the following function

$$f(x,y,z) = 10/(x^2 + y^2 + z^2 + a)$$

Over the region $-\infty \le x \le 0, -100 \le y \le 0$, and $-100 \le z \le 0$

Solution:

```
a = 2;
f = @(x,y,z) 10./(x.^2 + y.^2 + z.^2 + a);
format long
q1 = integral3(f,-Inf,0,-100,0,-100,0)
q2 = integral3(f,-Inf,0,-100,0,-100,0,'AbsTol',
0,'RelTol',1e-9)
```

Output:

q1 =

2.734244598320928e+03

q2 =

2.734244599944285e+03

20. Consider the function

$$f(x,y) = 1/(1 + x + y)$$

Integrate Over the triangular region bounded by

$$0 \le x \le 1, 0 \le y \le x.$$

Solution:

```
fun = @(x,y) 1./(1 + x + y);
```

```
ymax = @(x) x;
q = integral2(fun,0,1,0,ymax)
```

Output:

q =

0.261624071883185

References

[1] The MathWorks Inc. *MATLAB 7.0 (R14SP2).* The MathWorks Inc., 2005. The MathWorks Inc. MATLAB 7.0 (R14SP2). The MathWorks Inc., 2005.

[2] D. Houcque. Introduction to MATLAB for Engineering Students. *Northwestern University*, 2005.

Index

A

Absolute, 3
Acceleration, 1, 106
Access, 167, 184
Accurate, xi
Acronym, 167
Addition, v, 5, 6, 9, 168
Adjacent, 38
Adjoining, 58
Advance, 66
Aircraft, 55
Algebraic, 29, 37
Analogous, 73
Analysis, 1
Analytic, ix
Angle, vi, ix, 12, 29
Angled, 29
Angles, 12
Angular, 89
Applied, 46, 48, 49, 50
Arbitrary, 1, 92, 93, 95
Arc Length, vi, 57, 61
Arc rate, xi, 65
Area, vii, 1, 39, 122, 128
Arithmetic, x, 167
Arrow, 1
Atmosphere, 56
Axis, 2, 4, 5, 111, 137

B

Basic, v, 45, 55, 167, 168
Binormal, 60, 65
Bold, 1
Boundary, 135, 144, 153
Bounded, xii, 124, 129, 134

C

Cables, 29
Calculus, vi, 111
Capabilities, 167
Cartesian, 2, 4, 58
Centrifugal, 1
Charge, 1
Circle, 64, 117, 126, 143
Circular, 127, 133, 160
Circulation, vii, 112, 117
Clockwise, 66
Closed, 112, 114, 123, 129
Coefficients, 120
Collinear, v, 8
Command, 167, 168, 171, 187
Commands, x, 167
Common, 8
Commutative, 10, 30
Component, 71, 124, 128, 137
Components, v, 2, 5, 13, 25
Computations, 167
Concept, ix, 1, 7, 55
Cone, 76
Consecutive, 38, 60, 175, 176
Conservative, 91, 113, 114, 120
Constant, 48, 54, 56, 72, 74

Construction, 55
Contemporary, 167
Coordinate, 2, 4, 58, 71
Coplanar, v, 8, 107
Cosine, 12
Cosines, vi, ix, 12, 16
Counterbalance, 25
Critical, 55
Cross, 37, 38, 39, 60
Cube, 156, 157, 159, 170
Curl, 55, 73, 86, 89
Curvature, 60, 64, 67, 105
Curve, 67, 73, 82, 84
Curves, 57, 58, 84, 106
Cylinder, 129, 134

D

Debugging, 134
Decimal, 170, 172
Decreases, 75, 81
Density, 1, 56
Dependent, 1
Derivation, ix
Derivative, xii, 64, 71, 80
Derivatives, 64, 74, 75, 135
Difference, 10
Different, 1, 8, 10, 56, 167
Differentiable, 71, 86, 89, 146
Differential, vi, ix, 55, 73, 111
Differentiating, 147
Differentiation, vi, 111
Digit, 171
Digits, 170, 171
Dimensional, ix, 2, 3, 5, 57
Dimensions, 12, 55, 76
Directed, 1
Direction, ix, xi, 1, 2,
Directional, xii, 73, 74, 75
Directions, 7, 8, 10, 31
Displaced, 55
Displacement, 1, 46, 48, 51
Displacing, 112
Distance, 1, 74
Distributive, 10, 31, 40, 42
Divergence, vii, 55, 73, 86
Dot, xi, 29, 30, 40
Drag, 1
Dumped, 168

E

Elasticity, 55
Electric, 1, 45, 56, 87
Electrical, 1
Electromagnetic, ix, 103
Electromagnetism, 55, 112
Electrostatics, 55
Elements, 8, 31, 167, 178
Ellipse, 84, 127, 128, 129
Endpoints, 11, 58
Energy, 1
Environment, 167
Equal, 2, 10, 38, 59
Error, 158
Essentials, ix
Evaluation, vii, 123, 135
Execute, 167
Expression, 10, 40, 42, 115

F

Field, 1, 45, 55, 56, 73
Flowing, 56, 86, 123
Fluid flow, 3, 57, 88
Flux, 87, 123, 127, 128
Force, 1, 25, 26, 46
Forces, 25, 46, 48, 54
Frequency, 1

Function, 55, 57, 61, 73
Functions, 55, 56, 167, 158

G

Gauss, 154, 156, 159, 162
Generators, 55
Geometric, ix, xi, 1, 8
Geometrical, 1, 30
Geometrically, 30, 55
Geometry, ix, 45, 111
Gradient, 55, 7, 74, 81
Gravitational, 1, 56
Gravity, 25
Green, 135, 137, 138, 140

H

Hangs, 29
Heat flow, 55
Helix, 63
Hemisphere, 159
Hexagon, 14
High-performance, 167
Horizontal, 25

I

Incompressibility, 87
Incompressible, 87
Increasing, 66
Independence, ix, 113
Independent, vii, 113, 114, 115
Independently, 1
Initial, 1, 2, 7, 9
Input, 167, 159, 171, 175
Integral, 111, 122, 129, 135
Integrals, 111, 112, 113, 135
Integration, 111, 113, 114, 117
Intensity, 1, 56
Interpretation, vii, 38, 86, 69
Intersection, 56, 127, 144
Irrotational, 90, 92, 95, 112
Isothermal, 56

L

Laser, 55
Law, 8, 9, 10, 11
Length, 1, 3, 57, 59
Level curves, 84
Lift, 1
Line, vii, 8, 46, 58
Linear, 8, 89
Lines, 8, 85, 124, 136
Locus, 42

M

Machinery, ix
Magnetic, 1, 56
Magnitude, 1, 8, 18, 86
Magnitudes, 1, 50
Mass, 1, 86
Mathematical, ix, 167, 168
MATLAB, ix, 167, 170, 178
Matrices, 167
Maximum, 74, 108
Measure, vi, 30
Measured, 1, 74
Measurement, ix, 48
Messages, 167
Moduli, 10, 16, 31
Modulus, 3, 4, 9, 47
Moment, 1, 46, 47, 51
Momentum, 1
Mutually, 32, 52, 60

N

Negative, 2, 3, 6, 10
Neighboring, 60

Newline, 169
Normal, 60, 64, 74, 84
Null, 7, 39, 47

O

Octant, 124, 125, 126, 129
Operations, x, 167
Operator, vii, 73, 100
Opposite, 2, 6, 7, 10
Order, 60
Ordinary, 55, 73
Origin, 2, 11, 58, 117
Orthogonal, 66, 75, 81, 123
Osculating, 60, 64
Output, 168, 179, 186, 202
Outputs, x, 158
Outward, 86, 87, 146, 154

P

Paddlewheel, 90
Parabola, 114, 115, 144, 145
Parallel, 8, 39, 60, 86
Parallelepiped, 86, 87
Parallelogram, 8, 9, 13, 38
Parallelopiped, 87
Parameter, 57, 58, 61, 62
Parameters, 62
Parametric, 57, 67, 112, 117
Particle, 46, 48, 49, 51
Path, 112, 113, 115, 117
Perpendicular, 2, 31, 35, 47
Perpendiculars, 2
Physical, 1, 55, 86, 89
Physics, 45
Plane, 4, 47, 60, 122
Planes, 123, 129, 134, 155
Plot, 171, 173, 174, 201
Point, 1, 7, 47, 56
Points, 1, 2, 12, 60
Position, 2, 3, 12, 14
Positive, 6, 112, 126, 129
Potential, 1, 91, 93, 133
Power, 1, 167
Preserve, 170, 171
Principal, 60, 64
Print, 169, 175
Product, x, xi, 7, 200
Products, 29, 31, 45, 60
Program, 167, 168, 169, 179
Projecting, 148, 155
Projection, 30, 46, 73, 198
Projections, vi, 30

Q

Quantities, ix, 1, 55, 73
Quantity, 1, 46, 55

R

Radius, 56, 117, 128, 162
Rectangle, 140
Rectangular, 5, 152
Region, 55, 90, 112, 154
Regular, 14
Relativity, 111
Representation, ix, 1, 8, 112
Resultant, 9, 48, 49, 50
Revolution, 76
Right, 60, 66, 186
Robots, 55
Rotation, 64, 65, 89, 90
Rotational, 89

S

Scalar, ix, 1, 29, 46
Scalars, 1, 10, 55
Section, 6, 45, 64, 167

Segment, 1, 58
Segments, 8, 57
Semicolon, 168, 183, 186
Shuttle, 25
Sinks, 87
Skyscrapers, 25
Smoothly, 168
Solenoidal, 87, 98, 123
Sources, 87
Space, 1, 12, 25, 55
Special, 167, 183
Speed, 1
Sphere, 56, 125, 126, 159
Stationary, 56
Steady-state, 56
Steel, 25
Stokes, ix, vii, 135, 146
Straight, 58, 131, 132
Subset, ix
Subsurfaces, 122
Subtraction, 6, 168
Sum, 6, 8, 10, 178
Surface, 56, 74, 76, 108
Surfaces, 56, 74, 79, 122
Syntax, 168
System, 2, 4, 55, 60

T

Tangent, 58, 61, 64, 84
Tangents, 60, 72, 73
TaylorâĂŹs series, 86
Temperature, 1, 56
Tensions, 25
Terminal, 1, 7, 9, 12
Theorem, 113, 135, 145, 146
Thermo-dynamical, 55
Three, 2, 4, 12, 50
Thrust, 1
Time, 1, 56, 86, 123
Times, 5, 7, 175, 176
Tons, 25
Torsion, 64, 65, 69, 105
Totality, 56
Traced, 63
Transformation, 135
Transformed, 135
Triangle, 9, 10, 16, 124

U

Understand, 7, 12, 55, 79
Unit, 4, 5, 13, 18
Unity, 4, 5

V

Value, 1, 3, 19, 37
Variable, 55, 58, 115, 167
Variables, 115, 168
Vector, 1, 9, 18, 154
Vectors, 1, 5, 14, 60
Velocity, 1, 86, 89, 123
Vertex, 13
Vertical, 25, 125, 129, 145
Vertices, 13
Visualization, 167
Visualize, 55
Volume, 1, 87, 129, 163
Vortex, 90

W

Work, 1, 46, 48, 112
"Work done", 45, 48, 49, 122

Z

Zero, 4, 45, 82, 114

About the Authors

Nita H. Shah received her PhD in Statistics from Gujarat University in 1994. From February 1990 until now Professor Shah has been Head of the Department of Mathematics in Gujarat University, India. She is a post-doctoral visiting research fellow of University of New Brunswick, Canada. Professor Shah's research interests include inventory modeling in supply chains, robotic modeling, mathematical modeling of infectious diseases, image processing, dynamical systems and their applications, etc. She has published 13 monograph, 5 textbooks, and 475+ peer-reviewed research papers. Four edited books have been prepared for IGI-global and Springer with coeditor Dr. Mandeep Mittal. Her papers are published in high-impact journals such as those published by Elsevier, Interscience, and Taylor and Francis. According to Google scholar, the total number of citations is over 3334 and the maximum number of citations for a single paper is over 177. She has guided 28 PhD Students and 15 MPhil students. She has given talks in USA, Singapore, Canada, South Africa, Malaysia, and Indonesia. She was Vice-President of the Operational Research Society of India. She is Vice-President of the Association of Inventory Academia and Practitioner and a council member of the Indian Mathematical Society.

Jitendra Panchal is an Assistant Professor in the Department of Applied Sciences and Humanities, Parul University, India. He has 7+ years of teaching experience and 5+ years of research experience. His research interests are in the fields of mathematical control theory applied to various types of impulsive and/or fractional differential inclusions/systems with non-local conditions, and integer/fractional-order mathematical modeling of dynamical systems based on real-life phenomena. His 10 articles have been published in international journals indexed in SCIE, Scopus and Web of Science.